Fracture Analysis of the Composite Rock,

Numerical and Experimental

Using Brazilian Tests with

the Cement/Gypsum Specimen

Chia-Huei, Tu, Ph.D.

Professor Chao-Shi, Chen, Ph.D.

Department of Resources Engineering

National Cheng Kung University, Taiwan

To my beautiful and most precious family:
My parents Ke Tu and Shu-Sang Chen, my grandmother Ben Chou
My brothers Young-Huei and Cheng-Huei
Lovely girl Luma Yun-Ya Chan and her family

Chia-Huei Tu

To my wonderful wife Lily and our son Shen-Po

Chao-Shi Chen

Preface

Fracture mechanics has been suggested as a possible tool for solving a variety of rock engineering problems, such as rock cutting, hydro-fracturing, explosive fracturing, rock stability, etc. Rock linear elastic fracture mechanics is essentially based on the extension of Griffith theory and Irwin's modification to that theory which recognizes the importance of stress intensity near a crack tip. Irwin introduced parameters, called Stress Intensity Factors (SIFs), to express the stress and displacement field near a crack tip. In general, three SIFs, called K_I, K_{II} and K_{III} are introduced corresponding to three basic fracture modes, e.g. mode I (opening mode), mode II (sliding mode) and mode III (tearing mode), respectively. A superposition of the three modes describes the general case of loading (mixed mode loading). For a given body, the SIFs are known and the stresses and displacements near the crack tip can accordingly be determined. Therefore the problem of linear elastic fracture mechanics (LEFM) reduces to the determination of the crack tip SIFs. Besides, it is known that failure of rock construction frequently takes place following growth. Understanding the behavior of crack initiation and propagation is important for evaluating the safety limits of cracked structures. Crack propagation processes are frequently simulated by incremental crack extension analysis, based on certain failure criterion to predict the direction of crack initiation. For each increment of crack extension, a stress analysis is carried out at the crack tip, and the SIFs are evaluated. Because of the complex geometry, which is continuously changing during crack extension numerical techniques are required to simulate crack propagation and predict its path. Therefore, calculating the Stress Intensity Factors of Cracked Straight Through Brazilian Discs, and studying the path of crack propagation for each specimen, are the two main objectives of this book.

In order to help the reader to better understand the material dealt with in this book, the fundamentals of fracture mechanics have been summarized. In addition, the major research works already carried on the subject have been explained. Knowledgeable readers can bypass the first two chapters, and directly start by over-viewing the experimental investigation, as well as the results and their analysis.

The author wishes to thank the following professors for their reviews, observations, and input: Dr. T.H. Huang, National Taiwan University; Dr. K.J. Shou, National Chung Hsing University; Drs. D.H. Lee, and C.L. Wang, National Cheng Kung University.

November 20, 2010

Contents

Chapter 1

Introduction

Fracture mechanics is the field of mechanics concerned with the study of the formation of cracks in materials. Experts at an early stage of this theory were applied to metal and ceramic materials, and the results are satisfactory. In recent years, fracture mechanics has been widely used in rock engineering, the discussion of rock fracture mechanics related to more and more specialized books (e.g. Atkinson et al., 1987; Gramberg et al., 1989; Whittaker et al., 1992; Aliabadi, 1999), which include the major applications of rock drilling and excavation of the tunnel, the stability of oil drilling, hydro-fracturing and rock stability engineering etc.

Over the past few years, a considerable number of studies have been made on the homogeneous and isotropic media of the bi-material of fracture mechanics. Surprisingly, few studies have so far been made at anisotropic media. However, the natural rocks under the effects of the generated environment with clear bedding, foliation, layering, stratification, fissuring, schistosity or jointing. In general, these rock materials have properties that vary with direction and are said to be inherently anisotropic. Rock anisotropy is an important role in various engineering activities. In civil and mining engineering, rock anisotropy affects the stability of underground excavations, surface excavations, and foundations. Rock anisotropy affects drilling, blasting, and rock cutting. It can also induce directional fluid flow and contaminant transport. In petroleum engineering, rock anisotropy is a critical factor in controlling borehole deviation, stability, deformation, and failure. It also impacts fracturing and fracture propagation. Therefore, the degree of rock anisotropy study of rock fracture mechanics related factors must be considered.

1

This book describes formulations and computations of the BEM in transversely isotropic bi-material rocks. The work concentrates on the development of the BEM using the FORTRAN program, which are then applied to the two dimensional problems. This book is divided into six chapters. Chapter one defines the problem, introduce the undertaking of the study and outlines the method of approach adopted in this book. Chapter two provides an overview of linear elastic fracture mechanics studies of bi-material systems. Various solution techniques are introduced and studies concerning: (i) a crack lying along the interface, (ii) a crack terminating or crossing through an interface, and (iii) a wedge with its vertex on the interface are reviewed. In addition, a brief review of the numerical solution techniques, crack initial angle and the propagation path of the relevant literature. Chapter three provides a detailed account of the theoretical approach used to determine the stress and displacement fields using the boundary element method. This chapter includes the basic equations and fundamental solution of anisotropic elasticity, boundary element formulation, numerical discretization, stress intensity factor expression and the fracture propagation simulation. Chapter four proposes the determination of mixed mode stress intensity factors with the boundary element formulation. Numerical examples for determining the mixed mode stress intensity factors for several cracked materials are presented for isotropic and anisotropic media. Chapter five shows the experimental result of actual rocks and discussion, which includes the numerical results of the comparison with failure mechanism, the initial angle and the propagation path. Finally, Chapter six summarizes the findings and contribution of the current work.

Chapter 2

Literature Review

In this chapter, we provide an overview of linear elastic fracture mechanics (LEFM) studies concerning composite materials and structures. We begin with a brief description of major historical developments in fracture mechanics, and the LEFM approach adopted in treating cracks in homogeneous materials. This is followed by reviewing the most commonly adopted techniques for treating crack problems in bi-materials. Finally, studies concerning the failure of bi-material systems are reviewed and three categories are outlined: (i) a crack lying along an interface, (ii) a crack terminating or crossing an interface, and (iii) a wedge with its vertex on the interface. Both theoretical and experimental work are presented and discussed.

2.1 Major historical developments in Fracture Mechanics

The following Table 2.1 shows the evolution of fracture mechanics through research. The details of each contribution are developed below.

Table 2.1: Major historical developments in fracture mechanics

Problem	Coord. Syst.	Real/Complex	Solution	Date
Circular Hole	Polar	Real	Kirsch	1898
Elliptical Hole	Curvilinear	Complex	Inglis	1913
Crack	Cartesian	Complex	Westergaard	1939
V Notch	Polar	Complex	Williams	1952
Dissimilar Mar.	Polar	Complex	Williams	1958
Anisotropic Mat.	Cartesian	Complex	Sih	1965

2.1.1 Kirsch

In 1898, Kirsch showed that a Stress Concentration Factor (SCF) of three was found to exist around a circular hole in an infinite plate subjected to uniform tensile stresses. (Figure 2.1)

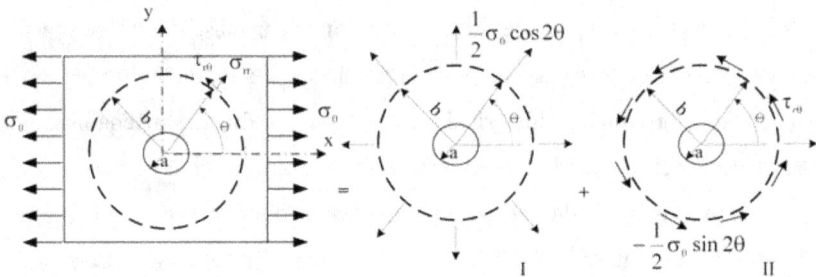

Figure 2.1: The model of Kirsch: circular hole in an infinite plate.

In this particular case, the state of stress can be expressed as follows:

$$\sigma_{rr} = \frac{\sigma_0}{2}\left(1 - \frac{a^2}{r^2}\right) + \left(1 + 3\frac{a^4}{r^4} - 4\frac{a^2}{r^2}\right)\frac{1}{2}\sigma_0\cos 2\theta$$

$$\sigma_{\theta\theta} = \frac{\sigma_0}{2}\left(1 + \frac{a^2}{r^2}\right) - \left(1 + 3\frac{a^4}{r^4}\right)\frac{1}{2}\sigma_0\cos 2\theta \qquad (2.1)$$

$$\tau_{r\theta} = -\left(1 - 3\frac{a^4}{r^4} - 2\frac{a^2}{r^2}\right)\frac{1}{2}\sigma_0\sin 2\theta$$

When $r = 0$, we obtain:

$$\sigma_{rr} = \tau_{r\theta} = 0, \quad (\sigma_{\theta\theta})_{r=a} = \sigma_0(1 - 2\cos 2\theta) \qquad (2.2)$$

For $\theta = \pi/2$ and $3\pi/2$, the expression of $\sigma_{\theta\theta}$ gives a SCF equal to 3.

2.1.2 Inglis

In 1913, while investigating the unexpected failure of naval ships (Figure 2.2), Inglis extended the solution for stresses around a circular hole in an infinite plate

4

to the more general case of an ellipse. Inglis showed that a concentration factor

of: $S.C.F = 1 + 2 \left(\frac{a}{\rho}\right)^{1/2}$ (2.3)

prevails around the ellipse (where a is the half length of the major axis, and ρ is the radius of curvature).

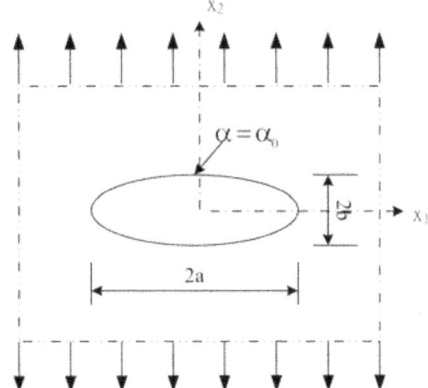

Figure 2.2: The model of Inglis, elliptical hole in an infinite plate.

Thus, at the tip of the ellipse, where $x_2 = 0$ and $x_1 = \pm a$, the expression of the stress becomes as follows:

$$\left(\sigma_{\beta\beta}\right)_{\alpha=\alpha 0}^{\beta=0.\pi} = \sigma_0 \left(1 + 2\sqrt{\frac{a}{\rho}}\right)$$ (2.4)

We notice that for $a = b$, we recover the stress concentration of 3 of a circular hole as found by Kirsch.

2.1.3 Griffith

In 1920, Griffith tested the strength of glass rods of different diameters at different temperatures and found that the strength increased rapidly as the size decreased. On the basis of this observation, Griffith's first major contribution to fracture mechanics was to suggest that internal minute flaws acted as stress raisers in solids, thus strongly affecting their tensile strength. Griffith determined, then, that the presences of minute elliptical flaws were responsible in dramatically reducing the glass strength from the theoretical value to the actually measured value. The second major contribution made by Griffith was in deriving a thermo-dynamical criterion for fracture by considering the total change taking place during cracking. During crack extension, potential energy is released and transferred to form surface energy.

2.1.4 Westergaard

In 1939, Westergaard derived an expression for the stress field near a sharp crack tip (Figure 2.3). The stress field near the crack tip was expressed as follows:

$$\sigma_{11} = \sigma_0 \sqrt{\frac{a}{2r}} \cos\frac{\theta}{2}\left(1 - \sin\frac{\theta}{2}\sin\frac{3\theta}{2}\right) + \cdots$$

$$\sigma_{22} = \sigma_0 \sqrt{\frac{a}{2r}} \cos\frac{\theta}{2}\left(1 + \sin\frac{\theta}{2}\sin\frac{3\theta}{2}\right) + \cdots$$

$$\sigma_{12} = \sigma_0 \sqrt{\frac{a}{2r}} \sin\frac{\theta}{2}\sin\frac{3\theta}{2}\cos\frac{\theta}{2} + \cdots \tag{2.5}$$

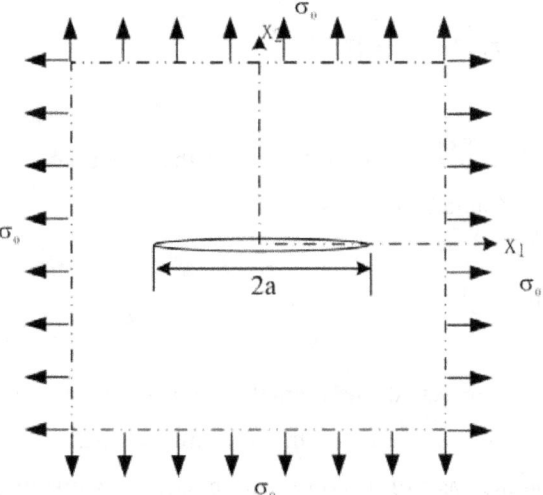

Figure 2.3: The model of Westergaard, stress field near a sharp crack tip.

William followed this work later, and extended this theory to cracks at the interface of two different homogeneous isotropic materials and made it applicable for V notches.

2.1.5 Irwin

In 1957, After World War II, Irwin made three major contributions to the fracture mechanics:

(i)   He extended Griffith's original theory to metals by accounting for yielding at the crack tip. This result is called the Modified Griffith's Theory.

(ii)   He altered Westergaard's general solution by introducing the concept of stress intensity factors (SIFs)

(iii)   He introduced the concept of energy release rate G.

Irwin introduced the concept of stress intensity factors defined as:

$$\begin{Bmatrix} K_I \\ K_{II} \\ K_{III} \end{Bmatrix} = lim_{r \to 0. \theta \to 0} \sqrt{2\pi r} \begin{Bmatrix} \sigma_{22} \\ \sigma_{12} \\ \sigma_{23} \end{Bmatrix} \qquad (2.6)$$

where σ_{ij} are the near crack tiop stresses and K_i are associated with three independent cinematic movements of the upper and lower crack surfaces with respect to each other as shown in Figure 2.4:

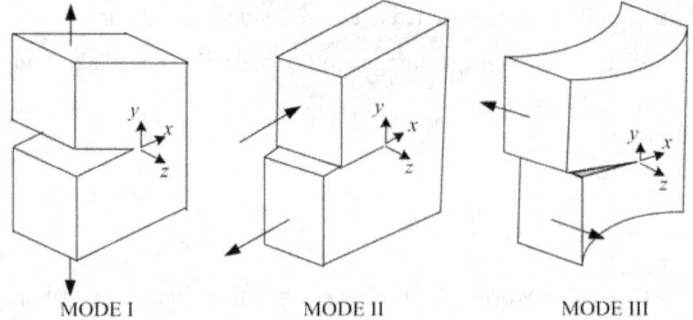

MODE I   MODE II   MODE III

Figure 2.4: The theory of Irwin, three main fracture modes.

The definitions of the three main fracture modes are explained as follow:

(i)   Opening Mode I: The two crack surfaces are pulled apart in the y-direction, but the deformations are symmetric about the x-z and x-y planes.

(ii)   Shearing Mode II: The two crack surfaces slide over each other in the x- direction, but the deformations are symmetric about x-y plane and skew symmetric about x-z planes.

(iii)   Tearing Mode III: The crack surfaces slide over each other in the z-direction, but the deformations are skew symmetric about the x-y and x-z planes.

With $\theta = 0$, and r the length of a small vector extending directly forward from the crack tip, we have:

$$K_I = \sqrt{2\pi r}\,\sigma_{22}\sqrt{\frac{a}{2r}} = \sigma\sqrt{\pi a} \qquad (2.7)$$

2.1.6 Williams

In the seminal paper by Williams (1952), he studied wedge problems

8

(V-notch in an angular plate) via the Eigenfunction expansion method, and found that unbounded stresses occur at the vertex for particular vertex angles. The asymptotic character of elastic stresses in wedge problem was therefore identified. (Figure 2.5)

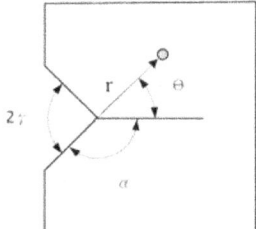

Figure 2.5: William's model, V-notch tip in an angular plate.

Later, in 1958, Williams discussed the singular behaviors of the ultimate wedge problem (i.e., cracks, as shown in Figure 2.6) in homogeneous and dissimilar media, respectively. The inverse square root singularity was recovered in homogeneous media, but the singularity with multipliers that result in oscillating behavior was found in dissimilar media. Williams proposed a method to determine the stresses and displacement at the wedge vertex (i.e., crack tip) of a crack at the interface between two dissimilar materials.

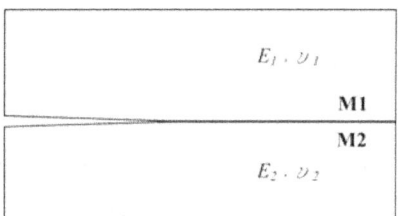

Figure 2.6: Crack at the interface between two dissimilar materials.

2.2 LEFM: The Stress Intensity Factors

The stress field in the vicinity of a crack tip in a homogeneous solid can be expressed by the following form of the stress tensor σ_{ij} (Meguid, 1989)

9

$$\sigma_{ij} = \frac{1}{\sqrt{r}} \left\{ K_I f_{ij}^I(\theta) + K_{II} f_{ij}^{II}(\theta) + K_{III} f_{ij}^{III}(\theta) \right\} \qquad (2.8)$$

where r, θ are the cylindrical polar coordinates defining the position of an element ahead of the crack-tip, as depicted in Figure 2.7. K_I, K_{II} and K_{III} are denoted the stress intensity factors (SIF) corresponding to the three basic modes of crack surface displacement. These modes, which are shown in Figure 2.4, are known as: (i) the opening mode (K_I) where the crack surfaces move directly apart, (ii) the sliding or in plane shear mode (K_{II}) where the crack surfaces slide relative to each other in a direction perpendicular to the leading edge of the crack, and (iii) the tearing or anti-plane shear mode (K_{III}) where the crack surfaces move relatively to one another parallel to the leading edge of the crack. The superposition of these three modes is sufficient to describe the most general case of crack surface displacement.

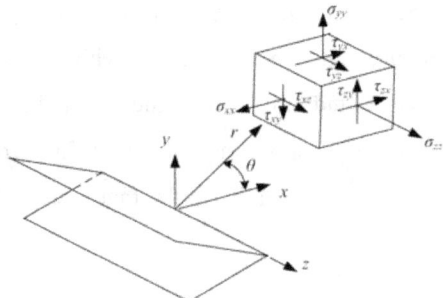

Figure 2.7: Crack-tip co-ordinates and stress state.

For the sake of simplicity, let us focus our attention to mode I loading where the stresses at the crack tip are given as:

$$\sigma_{xx}(r,\theta) = \frac{K_I}{\sqrt{2\pi r}} \cos\frac{\theta}{2} \left(1 - \sin\frac{\theta}{2} \sin\frac{3\theta}{2} \right)$$

$$\sigma_{yy}(r,\theta) = \frac{K_I}{\sqrt{2\pi r}} \cos\frac{\theta}{2} \left(1 + \sin\frac{\theta}{2} \sin\frac{3\theta}{2} \right) \qquad (2.9)$$

$$\tau_{xy}(r,\theta) = \frac{K_I}{\sqrt{2\pi r}} \cos\frac{\theta}{2} \sin\frac{\theta}{2} \cos\frac{3\theta}{2}$$

It is clear from the above expressions that the stress intensity factor K_I is an important parameter and is typically used to characterize the singular stress field around the crack tip. It connects with fracture theory through the postulate that fracture will commence (i.e. a crack will grow) when the stress intensity factor K_I reaches a critical value K_{IC}. This K_{IC} is an experimentally determined material property referred to as the fracture toughness of the material.

It should be noted, however, that a fracture criterion such as that described above is unambiguous only if the component is subjected to mode I loading. For mixed mode loading, however, a variety of criteria have been proposed for predicting the crack growth directions and critical loading conditions. Amongst those, the following three are the most commonly used: (i) the maximum tangential stress (*MTS*) criterion by Erdogan and Sih (1963), (ii) the maximum strain energy release rate (*G*) criterion, which was originally proposed by Griffith (1920) and modified later by Palaniswamy and Knauss (1978), Wu (1978) and others, and (iii) the minimum strain energy density (*SED*) criterion by Sih (1973).

2.2.1 Mixed mode

If a crack is under both tensile (mode I) and shear (mode II or III) loading conditions, this kind of mixed mode loading leads to mixed mode cracking. In LEFM, when two or three models are present simultaneously, the corresponding stresses and displacements for each mode described above can be added together to obtain the stresses and displacements for mixed mode. It should be noted that in the above expressions, only the singular terms are considered and higher-order terms have been neglected. Hence, they are applicable when $r \ll a$ (Ku, 2001).

11

Until now, most previous studies on rock fracture mechanics were mainly focused on mode I fracture (opening mode). However, data on mode II and mixed-mode fractures are scarce in the literature. Pre-existing cracks in a rock seldom subjects to tensile loading, but in the compressive, shear or mixed mode loading. For a given cracked body under a certain type of loading, the SIFs are known and the stresses and displacements near the crack tip can accordingly be determined. Hence, the problem of LEFM reduces to determine the SIFs of the crack tip.

Whittaker et al. (1992) concluded that the following three categories of methods are commonly used in determining the SIFs: (i) Analytical methods including the complex stress function, weight function, and stress concentration methods;(ii) Numerical methods including the finite element method (FEM), boundary element method (BEM), finite difference method (FDM), and boundary collocation method (BCM), and;(iii) Experimental methods including the commonly used photo-elastic techniques, acoustic emission and indirect measurements such as compliance calibration from laboratory tests.

The analytical solutions for the SIFs are limited to a small number of idealized situations in which the domain is considered to be infinitely. In practical situations the problem of interest is of finite size with complex loading. Therefore, there has been a need to develop numerical solutions using various methods such as boundary collocation method, the FEM, and the BEM.

2.3 Solution techniques for bi-material media containing cracks

A variety of analytical techniques have been used for treating bi-material solids containing cracks, among which are the Eigenfunction Expansion method, Mellin Transform method, Integral Equation method and Complex Function method. Numerical methods such as the boundary element method are also widely used.

2.3.1 Eigenfunction expansion method

The eigenfunction expansion method was developed by Williams in 1952 for treating elastic solids with sharp comers. Most notably, Williams used this method in 1959 to study the singular stress field of an interfacial crack in a bi-material and observed for the first time the oscillatory behavior. Other applications of the method can be found in (Williams, 1957, Zak and Williams, 1963, Swenson and Rau, 1970, and Pageau, et al., 1996). In the eigenfunction expansion method, Airy's stress function for each individual region (homogeneous, isotropic and defect-free), which satisfies the following bi-harmonic equation

$$\nabla^4 \phi(r, \theta) = 0, \tag{2.10}$$

is assumed to take the following form

$$\phi(r, \theta) = r^{\lambda+1} F(\theta)$$
$$= r^{\lambda+1}[a \sin(\lambda + 1)\theta + b \cos(\lambda + 1)\theta + c \sin(\lambda - 1)\theta + d \cos(\lambda - 1)\theta] \quad . \tag{2.11}$$

Since $\quad \sigma_r = \dfrac{1}{r^2}\dfrac{\partial^2 \phi}{\partial \theta^2} + \dfrac{1}{r}\dfrac{\partial \phi}{\partial r}$ $\hspace{3cm}$ (2.12)

$$\sigma_\theta = \frac{\partial^2 \phi}{\partial r^2} \tag{2.13}$$

$$\tau_{r\theta} = -\frac{1}{r}\frac{\partial^2 \phi}{\partial r \partial \theta} + \frac{1}{r^2}\frac{\partial^2 \phi}{\partial^2 \theta} \tag{2.14}$$

$$\frac{\partial u_r}{\partial r} = \frac{1}{2\mu}\left[\frac{1}{r}\frac{\partial \phi}{\partial r} + \frac{1}{r^2}\frac{\partial^2 \phi}{\partial^2 \theta} - (1 - \kappa)\nabla^2 \phi\right] \tag{2.15}$$

$$\frac{\partial u_\theta}{\partial r} - \frac{u_\theta}{r} + \frac{\partial u_r}{\partial \theta} = \frac{1}{\mu}\left(-\frac{1}{r}\frac{\partial^2 \phi}{\partial r \partial \theta} + \frac{1}{r^2}\frac{\partial \phi}{\partial \theta}\right) \tag{2.16}$$

the stresses and the displacements can be expressed in terms of λ and the coefficient a, b, c and d. Enforcing the appropriate boundary conditions, which include the traction-free conditions along the free surfaces and the continuity conditions along the interface, we are able to derive the so-called eigenequations:

$$[E(\lambda)]\{X\} = \{0\} \tag{2.17}$$

from which the eigenvalue λ and the associated eigenvector $\{X\} = \{a, b, c, d\}^T$ can be determined. The singular eigenvalue which satisfies $0 < Re(\lambda) < 1$

determines the stress singularity ($\zeta = 1 - \lambda$), while the eigenvector {X} defines the angular variation functions $f_{ij}(\theta)$. This method can be used for 2-D problems as well as 3-D problems. But it cannot give information regarding the stress intensity factors.

2.3.2 Mellin transforms method

According to the Mellin transform of a function f (r) is defined as: (Bogy, 1968)

$$F(s) = \int_0^\infty f(r)\, r^{s-1}\, dr \qquad (2.18)$$

and its inversion is given by :

$$f(r) = \frac{1}{2\pi i} \int_{c-i\infty}^{c+i\infty} F(s)\, r^{-s}\, ds. \qquad (2.19)$$

To solve a crack problem in a bimaterial system, Equation (2.12) - (2.16), the boundary and the loading conditions are described in terms of the transform domain (s) by using the Mellin transform (Equation (2.18)). A characteristic equation is then derived and the power s can thus be obtained. In this case, die stress singularity is given by $\zeta = -s - 1$, where $-2 < Re(s) < -1$.

This method can be used in 2-D problems for determining both stress singularities and stress intensity factors. Nevertheless, it has, so far, been mainly used for determining the stress singularities (Bogy, 1968, 1971, Bogy and Wang, 1971, and Wijeyewickrema, et al., 1995).

2.3.3 Integral equations

Using the method of integral equations, a crack problem is formulated in terms of a system of singular integral equation(s). Separating the original problem into a homogeneous problem without flaws and a stress disturbance problem due to the existence of a flaw or crack, the second problem can usually

14

be formulated in terms of an integral equation of the following form:

$$A\phi(x) + \frac{1}{\pi}\int_{-a}^{a}\left[\frac{B}{t-x} + k_s(x,t) + k_f(x,t)\right]\phi(t)\,dt = g(x), (-a < x < a) \quad (2.20)$$

where the constant A and B and the functions ϕ, g, k_s and k_f may be scalar (symmetric problems) or matrix quantities (non-symmetric problems). The unknown function ϕ is defined by a dislocation density function of the form:

$$\phi(x) = \frac{\partial}{\partial x}\left(u^+(x) - u^-(x)\right) \quad\quad\quad (2.21)$$

where x is the coordinate along the crack and $u^+(s)$ and $u^-(x)$ are the crack surface displacement vectors on the opposite faces of the crack. The known function g(x) corresponds to the crack surface tractions. $k_f(x, t)$ is a Fredholm kernel which is square integrable in the closed domain {-a, a}. The kernel $k_s(x, t)$, on the other hand, is a characteristic feature of the problems which involve cracks intersecting a bi-material interface or a free surface. Equation (2.20) may be obtained either by using the dislocation solutions (if available) as Green's functions or by applying integral transformations.

Integral equations of the form given by Equation (2.20) can seldom be solved in a closed form. However, by applying the residue theorem, the singularity of the solution can be obtained. The problem can then be solved within sufficient accuracy by incorporating this singularity into the numerical solution of the integral equation. Details of this method can be found in (Cook and Erdogan, 1972, Erdogan and Biricikoglu, 1973, and Erdogan et al., 1973).

2.3.4 Numerical methods

The finite element method (FEM) has been widely used in fracture mechanics studies. Due to the singular behavior near the crack tip, conventional finite elements are generally incapable of correctly describing the stress field. To enhance the accuracy of the numerical solutions without significantly increasing the mesh density, quarter-point elements were proposed by Henshell and Shaw (1975) and Barsoum (1976) for cracks in homogeneous media. Finite elements with embedded singularities were also proposed by Lin and Mar (1976) and

Meguid et al. (Meguid et al., 1995, and Tan and Meguid, 1996) for bi-material solids containing cracks.

Special boundary elements have also been developed for cracks at bi-material interfaces, e.g. by Kwon and Dutton (1991). More references regarding the application of the BEM in bi-material solids can be found in the review of Aliabadi (1997). These solution methods have been used to treat a wide variety of problems. The following sections provide a summary of some of the most relevant investigations.

Recently, the two most popular techniques in continuum mechanics, namely FEM and BEM, are used in the analysis of the practical engineering problems. The BEM is firmly established in many engineering disciplines as an effective alternative computational tool the FEM.

In recent years, the BEM is well-established numerical technique in the engineering problem. The BEM has proven to be a powerful numerical technique that has some advantages over the domain-based method such as finite element method. The advantages and disadvantages of the BEM are summarized as follows (Becker, 1992):

Advantages of the BEM

1. Less data preparation time.

2. High resolution of stresses.

3. Less computer time and storage.

4. Less unwanted information.

5. Easily applicable to incompressible materials.

Disadvantages of the BEM

1. Unfamiliar mathematics.

2. The interior must be modeled in non-linear problems.

The BEM is a numerical computational method of solving linear partial differential equations which have been formulated as integral equations (i.e. in

16

boundary integral form). It can be applied in many areas of engineering and science including fluid mechanics, acoustics, electromagnetic, fracture mechanics, and plasticity.

Recently, the BEM is used in the fracture analysis of structures and rocks, because of the complex shape and continuously changing path of the propagating crack. The main attraction of BEM is reduction in the dimensionality of the problem: for two-dimensional problems, only the line boundary of the domain needs to be discretized into elements. The simpler modeling of the body means that regions of high stress concentration can be modeled in more detail as the necessary high concentration of elements is confined to one less dimension. Another important feature of the boundary element formulation is that it provides a continuous modeling of the interior since no discretization of the interior is required; this leads to a high resolution of internal stresses and displacements.

2.4 Anisotropic behaviors of fractured rocks

Studies on the mechanical behaviors of anisotropic and fractured rocks are on the increase. These behaviors depend on the elastic properties of rock material and structural characteristics of rock mass: geological discontinuities and anisotropies, spatial attitude (strike and dip), their orientation relative to stress states, their frequency or spacing and their mechanical properties (Amadei et al, 1987; Savage et al., 1986). This may be attributed to the following factors: (i) the directional variation of elastic intact rock properties (E, v) in the rocks, especially if it is marked by anisotropy; (ii) the variation in compressibility of the fractured rocks related to joint closure: a function of surface roughness, the degree of interlocking and the magnitude of normal stress on the discontinuity surfaces (Goodman, 1976; Bandis et al., 1983); (iii) the spatial attitude and geometric configuration of the different joint sets within the rocks; (iv) the density of fracturing or the joint spacing in each set; (v) the mechanical properties of the joints, joint sets and the rocks. These are the factors causing a variation of the gravitational stresses in regularly jointed rocks. Other external factors such as interstitial pore water pressure in the rock mass and the tectonic stresses in the vicinity add to the complexity of the problem.

The elastic behaviors in anisotropic and jointed rocks receive less attention and only a few studies on this problem (Bray, 1967; John, 1970, Amadei et al., 1985; Savage et al., 1986; Brady et al., 1986; Liao, 1990; Pan, 1993; Chen, 1996). These studies show that the magnitude and orientation of the in situ stress field induced in a rock mass under gravity is strongly reliant on the geologic structures of the rock mass. Amadei et al. (1987) proposed models taking into account the effects of horizontal and vertical anisotropy (transversely isotropic) and regularly jointed rock masses modeled as an equivalent transversely isotropic elastic continuum under gravity and laterally restrained conditions. It is demonstrated that, under these conditions, the induced horizontal stress can be larger than, equal to or less than the vertical stress in anisotropic rocks. It is also found that the stress distribution is affected by the joints at shallow depth, and converges to empirical distributions or to the classical isotropic solution proposed in the literature as depth and/or joint spacing increases, or as rock mass anisotropy decreases.

Because rocks are largely discontinuous, anisotropic and inhomogeneous in natural geological state, difficulties arise in numerical modeling due to such complex and non-homogeneous geological conditions of rock mass. The complex combination of its constituents and its long history of formation make rock mass a difficult material for mathematical representation via numerical modeling. Rock mass can be generally classified into three groups, i.e. (A) continuous, (B) discontinuous and (C) pseudo continuous groups. Type (A) refers to intact rock mass, type (B) represents jointed rock mass and type (C) is for highly fractured or weathered rock mass. The behavior of type (A) rocks can be analyzed by means of model based on continuum mechanics, while a discontinuous model such as those proposed by Cundall (1971) may be used for analyzing the type (B) rock mass where joint elements in the finite element

18

analysis are also useful. Discontinuous model similar to that of type (B) can be used for type (C) rock mass. However, it is almost impossible to explore all the joint systems and it seems that this type of rock mass behaves just like a continuous body in a global sense. Therefore, a continuum mechanics model can be used with the effect of discontinuities adequately considered in the model. This is achieved by homogenization technique where equivalent continuum properties of the rock mass are derived based on the geometry of the contained fracture systems and physical properties of the intact rock matrix and the fractures.

The concepts of continuum and discontinue are not absolute but relative. The choice of continuum or discrete (discontinue) methods depends on many problem specific factors, and mainly on the problem scale and fracture system geometry. This is particularly true for rock mechanics problems. There are no absolute advantages of one method over another. Some of the disadvantages of each type can be avoided by using the continuum discrete models, termed hybrid models. The investigation of anisotropy of the samples was performed on the basis of mathematical modeling for anisotropic material of Lekhnitskii (1963) and Pan (1996a). The description of the symmetric anisotropy of the elastic properties of rocks and the assessment of their anisotropic material behavior according to the tectonic background and endogenous lateral compactions is based on the concepts of the theory of elasticity and uses the results of static and dynamic measuring procedures.

In the analytical approach, the symmetric anisotropy and the elastic properties are considered along the main direction of anisotropy (derived from ultrasonic investigations). The relationships between stress, deformation and the elastic constants of a rock body are described with respect to the main directions of a global Cartesian coordinate system (x, y and z) which coincides with the main tectonic deformation axes. The elastostatic theories and the basic relationships for the description of anisotropic and isotropic bodies are presented in the works of Lekhnitskii (1963), Liao (1990) and Amadei and Pan (1992). In the

19

orthogonal anisotropy case, there are three elasticity moduli (E_x, E_y and E_z), six Poisson's ratios (v_{yx}, v_{zx}, v_{xy}, v_{zy}, v_{xz}, v_{yz}) and three shear moduli (G_{yz}, G_{xz} and G_{xy}) corresponding to the main coordinate axes. In the transverse isotropy case, the system reduces to two elasticity moduli ($E_x = E_y$ and E_z), three Poisson's ratios ($v_{yx} = v_{xy}$, $v_{zx} = v_{zy}$ and $v_{xz} = v_{yz}$) and two shear moduli ($G_{yz} = G_{xz}$ and G_{xy}). For the description of isotropic systems, only two parameters (E and v) are required. For the Poisson's ratios, the first index denotes the direction of transversal strain and the second index, the direction of longitudinal length deformation. The calculation of the dynamic elasticity is based on measurements of ultrasonic travel time with respect to the axes of tectonic deformation in the samples.

For transversely isotropic rock mass with planes of elastic symmetry striking parallel to the z-axis and dipping an angle with respect to the flat ground surface (Liao, 1990; Amadei, 1996), there are 13 non-zero elastic compliance components in the global coordinate system (x, y, z). However, they can be expressed in terms of the five independent elastic constants (E, E', v, v' and G') defined in the local coordinate system (x', y' and z') and the dip angle. After coordinate transformation, the number of elastic compliances is changed from 5 in the local coordinate system to 13 in the global coordinate system. Also, if the strike of the layers in the rock mass is not parallel to the z-axis, the number of elastic compliances could be 21 in the global coordinate system.

2.5 Analysis of cracks along an interface

In 1959, Williams examined the local stress field near the tip of a crack which lies along a bi-material interface and observed an oscillatory stress singularity of the form $r^{-\frac{1}{2}+i\eta}$ for traction free boundary conditions on the crack surfaces. In the early 60's, Sih and Rice (1964) and Rice and Sih (1965) provided explicit expressions for the near-tip stresses and related them to remote elastic fields. Erdogan (1963), England (1965), and Malyshev and Salganik (1965) further examined two-dimensional singular models for single and multiple crack

20

geometries. Their solutions revealed that the stress and displacement fields can be described as:

$$\sigma \sim r^{-\frac{1}{2}}(sin, cos)(\eta \, log \, r) \qquad\qquad (2.22)$$

$$u \sim r^{\frac{1}{2}}(sin, cos)(\eta \, log \, r) \qquad\qquad (2.23)$$

$$\eta = \frac{1}{2\pi} log\left(\frac{1-\beta^*}{1+\beta^*}\right) \qquad\qquad (2.24)$$

with β^* being a dimensionless composite parameter, which depends upon the material properties of the composite solid (Dundurs, 1969).

The above relations reveal the following difficulties: (i) rapid oscillations are present in the stress and a displacement field, (ii) the problem ceases to be self-similar, and (iii) modes I and II are inherently coupled. In addition, without specifying a characteristic length, the solution results in logarithmically infinite factors in the usual definition of the SIF (Comninou, 1990).

Various attempts have been made to overcome the above difficulties. Notably, the work of Comninou (1977) and Atkinson (1977) has received the attention of the scientific community. To rid the model of the oscillatory singularity, Comninou suggested a model which involves frictionless contact of the crack surfaces (Figure 2.8). With such a model, a contact zone length could be determined and, in principle, the stresses in the contact region could also be checked to ensure that they are compressive. Various extensions of the work were made by Comninou (1977) and co-workers (Comninou, 1978, Comninou and Schmueser, 1979, and Comninou and Dundurs, 1980) and the treatment was based on the numerical solution of the corresponding integral equations. Analytical treatments using this model were also given by Atkinson (1982, 1983) and Gautesen and Dundurs (1987, 1988). The contact model eliminated the problem of interpenetration of crack surfaces. However, it led to a surprising result; namely, that for materials which are only slightly different and loaded under simple tension, the interface crack tip has a zero K_I and nonzero K_{II}.

21

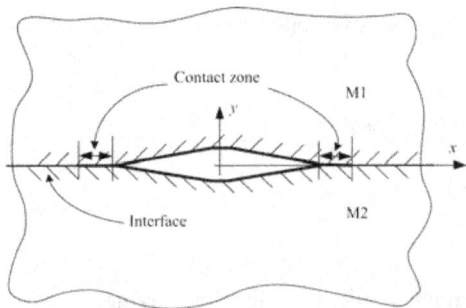

Figure 2.8: Comninou's contact model

Atkinson (1977), on the other hand, resolved the difficulties associated with the oscillatory singularities by introducing a transition layer between the two constituent materials. He proposed two models. In the first (Figure 2.9 (a)), he replaced the interface by a thin strip of finite thickness with a discontinuous interface. Accordingly, the crack is constrained to lie in the interior of the interface layer. Since there are still two discontinuous interfaces in this model, it did not gain much popularity. In the second model (Figure 2.9 (b)), the interface layer or connecting strip has continuously varying moduli which are equal to the surrounding solids at the boundaries. The resulting solutions avoided the oscillatory behavior. The singularity at the crack tip is of the usual square-root type. This model was also examined by Dale and Erdogan (1988) and Erdogan et al. (1991). However, this model dramatically increases the complexity for stress analysis (Atkinson and Craster, 1995). In addition, the solution depends to a large extent upon the thickness of the strip, relative position of the crack in the strip and the law governing the moduli variation.

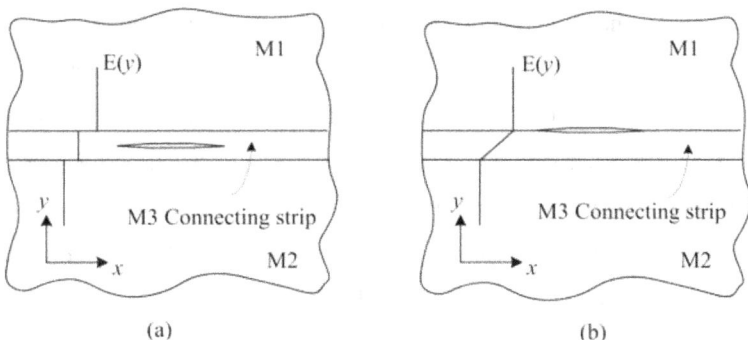

(a)   (b)

Figure 2.9: Atkinson's connecting strip models

Based on the fact that the predicted contact zone is extremely small compared with the crack length, Rice (1988) suggested the concept of "small scale contact" which essentially ignores the presence of crack surface interpenetration in solutions of William's type. Following Hutchinson et al., (1987), Rice introduced the notion of a complex stress intensity factor $Kr^{i\varepsilon} = K_I + iK_{II}$, with K_I and K_{II} being the stress intensity factors. He also clarified the applicability of the approach to various situations.

Aside from the stress field analysis, another important issue in fracture mechanics studies is the prediction of crack growth. He and Hutchinson (1989) investigated the initiation of kinking of an interfacial crack between two linear elastic solids. Using an asymptotic analysis, they expressed the stress intensity factors and the energy release rate of a kinked crack in terms of the corresponding quantities for the interfacial crack prior to kinking. By examining the resultant energy release rate, they predicted the kink directions corresponding to various material combinations. Geubelle and Knauss (1994) further examined the problem by applying both the maximum energy release rate (G) and the maximum tensile stress *(MTS)* criteria. They found that the kink direction cannot be uniquely determined without specifying a fixed crack extension length. An experimental study accompanying the above theoretical work was carried out by Bowen and Knauss (1993) using a bi-material specimen made of two elastomers. They observed that the crack may advance by kinking

23

into either of the constituent materials or propagating along the interface, depending on the applied loading conditions. By comparing the experimentally observed kink direction with the theoretical predictions given in (He and Hutchinson, 1989), they found that the numerical and experimental data do exhibit the same trend, but do not agree uniformly. They also found that the agreement can be improved by a suitable choice of a length parameter corresponding to the initial (virtual) crack extension.

When a bi-material solid containing an interfacial crack is subjected to dynamic loading, the problem becomes much more complicated. Because of that, investigations concerning the dynamic behavior of bi-material solids have been meager and most of the existing works are limited to infinite full or half spaces. Among those, Sih and Chen (1980) studied the impact response of a crack located symmetrically in the middle of a layer sandwiched between two half spaces. Their technique assumed several symmetrical conditions, and hence cannot be applied to the interfacial crack problem. Kuo (1984) carried out numerical and analytical studies of the transient response of an interfacial crack between two dissimilar orthotropic half-spaces. Yang and Bogy (1985) and Qu (1996) computed dynamic stress intensity factors of an interfacial crack in a layered half-space for inplane problems. The transient response of an interfacial crack in a layered plate under antiplane loading was also considered by Kundu (1985). To the best knowledge of the author, there has been no dynamic solution for interfacial cracks in finite solids under in-plane loading.

Due to the complexity of the problem, relatively few experimental investigations have been reported. Among those dealing with local stress and displacement fields near the interfacial crack tip, Liechti and Knauss (1982) studied debonding of a sandwiched elastomer in a uniformly loaded strip interferometrically. They observed extensive three-dimensional deformation in such specimens. Gdoutos (1982) employed the photoelastic method to investigate problems of composite materials having stress concentrations. Lu and Chiang (1993) also utilized photo-elasticity to examine the crack-tip stress field in a bi-material model made from two dissimilar photo-elastic materials. Following Rice's (1988) definition of the stress intensity factors, they evaluated the stress intensity factors which in general agree with analytical solutions. Interestingly, they observed that the fringe loop inclination angle θ_m, which is proportional to K_{II}/K_I, varies from one loop to another. This phenomenon illustrates in essence the rotational characteristics of the singular stress field or in other words "the oscillatory nature" of the stress singularity. Using the Coherent Gradient Sensing method, Tippur and Rosakis (1991) examined the crack-tip deformation fields in quasistatically as well as dynamically growing cracks along bi-material interfaces. To study the problem of interfacial crack propagation during surface-layer removal by scraping, Miskioglu et al., (1991) used photo-elastic technique to determine the direction of maximum strain energy release rate.

2.6 Analysis of cracks terminating at an interface

Zak and Williams (1963) extended Williams' (1952) eigenfunction expansion method to obtain a solution for two dissimilar elastic half spaces; one of which contained a semi-infinite crack perpendicular to the interface (Figure 2.10(a)). Their analysis revealed that the stress singularity is of the order $r^{-\xi}$, where ξ is real but dependent on the elastic moduli of the bonded materials. Swenson and Rau (1970) studied the same geometry as Zak and Williams (1963), but under plane strain conditions. In addition to the effect of the elastic mismatch on the power of the stress singularity, they also obtained the relative magnitudes of the various stress components under uniaxial loading conditions. Based on these relative magnitudes, the possible failure modes of the composite solid were discussed.

To obtain the stress intensity factor of a crack which is perpendicular to the interface of two bonded half-spaces and subjected to concentrated wedge loading, Cook and Erdogan (1972) used the Mellin transform method to formulate the problem and to derive the appropriate integral equations necessary to describe the problem. Erdogan and Biricikoglu (1973) further studied the problem of two bonded half-spaces containing a finite crack perpendicular to and crossing through the interface (Figure 2.10(b)). The problem was formulated as a system of singular integral equations with generalized Cauchy kernel, which enabled the numerical determination of the stress intensity factors. For two bonded half-spaces and under an additional assumption that the crack geometries and loading conditions have a plane of symmetry perpendicular to the interface, Lu and Erdogan (1983) examined the singular behavior of the T-shaped crack (Figure 2.10(c)) and the cross-shaped crack (Figure 2.10(d)).

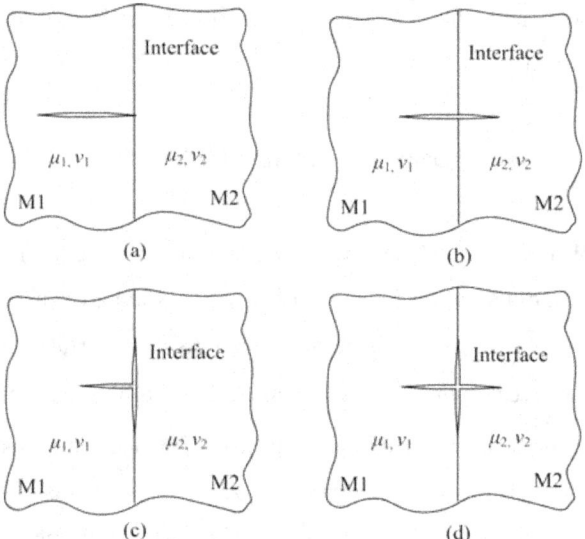

Figure 2.10: Crack perpendicular to an interface

For treating the problem of a finite elastic body containing a crack perpendicular to the bi-material interface, Meguid et al. (1995) presented a new singular finite element in which the proper singularities are embedded. The effects of elastic mismatch and crack configuration upon the resulting stress intensity factors were evaluated. This approach was further extended to evaluate the dynamic stress intensity factors (Tan and Meguid, 1996).

Bogy (1971) extended the Mellin transform method to investigate the dependence of the order of stress singularity on the material constants and the crack angle when a semi-infinite crack terminates at an arbitrary angle to the interface. Bogy's solution indicates that for angles which are nearly 90 degrees to the interface, the singularity ζ is a real number; for others, ζ takes a complex value. Wijeyewickrema, et al. (1995) further studied the problem of a crack terminating at a frictional interface between two materials using Mellin integral transform. The characteristic equations which yield the order of the crack-tip singularity are obtained in terms of the Dundurs constants, the inclination of the crack and the coefficient of friction. For the special case when the crack is perpendicular to the interface and the two wedges slip in opposite directions, it was shown that the problem decouples into mode I (symmetric) and mode II (anti-symmetric) cracks. In a recent article, Tan and Meguid (1997) considered a more general geometric configuration of a sharp notch terminating at a bi-material interface. Employing the complex function method, the characteristics of the singular stress field were described and, the dependence of the stress singularities upon the elastic mismatch and notch geometry was demonstrated.

To the best of the author's knowledge, there have been only two articles dealing with dynamic analysis of cracks terminating at bi-material interfaces, one by Atkinson (1977) and the other by Tan and Meguid (1996). Both of the works examined cracks perpendicular to bi-material interfaces. In (Atkinson, 1977), an analytical method was used to treat the anti-plane problem, while in (Tan and Meguid, 1996), a singular finite element was employed to analyze the in-plane problem.

On the experimental side, Wang and Chen (1993) used photo-elasticity to determine the stress distribution and the SIFs of a crack perpendicular to an interface. Their results showed that the far-field effects play a significant role in the resulting stress distribution and SIFs solutions. This technique was further used by Chen and Wang (1997) for examining the singular stress field of a crack terminating at an arbitrary angle to the bi-material interface.

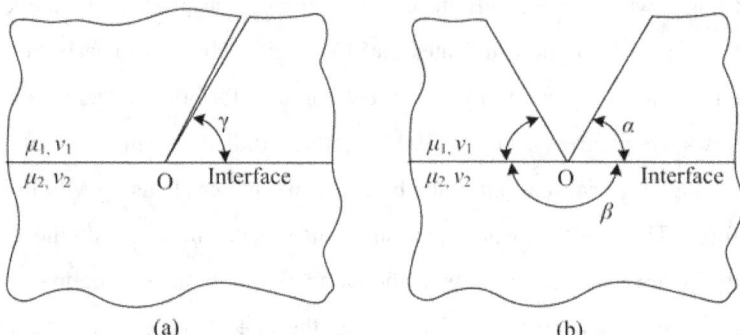

(a)   (b)

Figure 2.11: (a) An oblique crack terminating at an interface, and
(b) a sharp notch terminating at an interface

2.7 Crack propagation path in numerical simulation

Crack propagation in many brittle materials can take place when the crack is subjected to pure mode I and mixed mode I-II loading. To determine the SIF for pure mode I, Libatskii and Kovchik (1967) used an approximate integral solution, Rooke and Tweed (1973) the Fredholm equation, Isida (1975) the boundary collocation method, Murakami (1976) the finite element method, Guo et al. (1993) an analytical expression to an infinite cracked plate, and Fischer et al. (1996) the finite element method combined with the modified ring test.

By varying the orientation of the crack relative to the loading direction, Gandhi (1972) used the modified mapping-collocation method for orthotropic rectangular plates, Awaji and Sato (1978) the dislocation and boundary collocation methods with the superposition procedure for a Brazilian disc with a central crack, Atkinson et al. (1982) a distributed dislocation method for cracked Brazilian disc, Lim and Johnston (1993) the finite element method for semicircular specimen under three-point bending loading, Fowell and Xu (1994) the dislocation method combined with the superposition technique for Cracked Straight Through Brazilian Disc (CSTBD), Wang et al. (2003) the sub-model method in finite element analysis software-ANSYS for cracked chevron notched Brazilian disc (CCNBD), and Dong et al. (2004) the weight function method for Brazilian disc with a central crack. All the above references concur that the geometry of the Brazilian disc has a number of advantages, i.e. simpler specimen preparation, higher failure load and easier testing procedure, over other available methods in rock fracture tests. However, all the conventional analysis methods are limited to isotropic media except for the work done by Chen et al. (1996) that proposed the BEM with J-integral to determine the pure mode I and pure mode II SIFs for cracked discs of anisotropic rock under diametral loading. The J-integral technique, though very accurate, has difficulty in handling the cases with consideration given to the body force, pressurized crack, and curved crack.

Crack propagation leading to rock failure is a very important topic in rock mechanics research. A number of studies have been done on two-dimensional model plates with through going pre-existing fractures and some of them have been done on three-dimensional specimens (Hoek and Bieniawski, 1965; Peng and Johnson, 1972; Hallbauer et al., 1973; Bobet, 2000; Sato et al., 2001; Wong et al., 2001; Al-Shayea, 2005). In reality pre-existing fractures are 3D in nature. The propagation mechanisms of a 3D crack may be more complicated. Actually, according to the observations by Germanovich et al. (1994) and Germanovich and Dvskin (2000), unlike in 2-D samples, there are intrinsic limits on the growth of a crack in a 3D model. However, the crack propagation of rocks containing 3-D cracks is out of the scope of the present study. In particular, the behaviors of crack propagation of 2-D model have not been fully investigated, and the influence of anisotropy on the growth of a crack is still not fully understood. Thus, the present study is focused only on the propagation of 2D cracks.

In 2D crack problems, stress intensity factors are important in analysis of cracked materials. They are directly related to crack propagation criteria. The singularity of stresses near a crack tip is a challenge to numerical modeling

29

methods, even to the BEM. Because the coincidence of the crack surfaces gives rise to a singular system of algebraic equations, the solution of cracked problem cannot be obtained with the direct formulation of the BEM. Several special methods within the scope of the BEM have been suggested for handling stress singularities, such as the Green's function method (Snyder and Cruse, 1975), the sub-regional method (Blandford et al., 1981; Sollero and Aliabadi, 1993; Sollero et al., 1994), the displacement discontinuity method (Crouch and Starfield, 1983; Shen and Stephansson, 1994; Scavia, 1995).

The Green's function method overcomes the crack modeling problem without considering any source point along the crack boundaries. This method has the advantage of avoiding crack surface modeling and gives excellent accuracy; it is, however, restricted to very simple crack geometries for which analytical Green's function are available. The sub-regional method has the advantage of modeling cracks with any geometric shape. The method has the disadvantage of introducing the artificial boundaries of the original region into several sub-regions, thus resulting in a large system of equations. In fracture propagation analysis, these artificial boundaries must be repeatedly introduced for each increment of the crack extension. Therefore, this method cannot be easily implemented as an automatic procedure in an incremental analysis of crack extension problems. The displacement discontinuity method overcomes the crack modeling difficultly by replacing each pair of coincident source points on crack boundaries by a single source point (Poterla, 1993). Instead of using the Green's stresses and displacements from point forces, the displacement discontinuity method uses Green's functions corresponding to point dislocations (i.e., displacement discontinuities). This method is quite suitable for crack problems in infinite domains where there is uncrack boundary. However, it alone may not be efficient for finite domain problems, since the kernel functions in displacement discontinuity method involve singularities with order higher than those in traditional displacement BEM. Hence, this method is not suitable for problems involving finite domains.

30

Recently, several single-domain BEMs have been proposed for the study of cracked media (Pan and Amadei, 1996; Pan, 1997; Aliabadi, 1997; Chen and Hong, 1999; Cisilino and Aliabadi, 1999). It involves two sets of boundary integral equations and is, in general, superior to the aforementioned BEM's. As a consequence, general mixed mode crack problems can be solved in a single-domain BEM formulation. The single-domain analysis can eliminate re-meshing problems, which are typical of the FEM and the sub-regional BEM. The single-domain BEM has received considerable attention and has been found to be a good method for simulating crack propagation processes.

One of the single-domain BEMs is the so-called Dual Boundary Element Method (DBEM). The essence of this technique is to apply displacement integral equations at one surface of a crack element and traction integral equation at opposite surface, although the two opposing surfaces occupy practically the same space in the model. (Jing, 2003) The hypersingularity involved in the traction integral is evaluated analytically by assuming a piece-wise flat crack path. The term DBEM was first presented in Portela (1993) and Poterla et al., (1993). Extension of this DBEM formulation to the 2-D anisotropic crack problem was reported in Sollero et al., (1994).

In the DBEM formulation, the displacement on each side of the crack surface is collocated as unknown. Thus, the resulting algebraic equations are doubled along the crack surface, which may be unnecessary for SIF calculation. Therefore, an ideal single-domain BEM formulation would be the one which requires discretization on one side of crack surface only. Such single-domain BEM formulation can be achieved by applying the displacement integral equation to the no-crack boundary only, and the traction integral equation on one side of the crack surface only. Since only one side of the crack surface is collocated, one needs to choose either the relative crack displacement (RCD). This new BEM formulation can be applied to the general fracture mechanics analysis in anisotropic media while keeping the single-domain merit.

Chapter 3

Theoretical Background

This chapter consists of four parts: the LEFM for anisotropic and fundamental Solution, the boundary element formulation, the calculation of stress intensity factor and the fracture propagation simulation. With an overview of the fundamental theories of the fracture mechanics of bi-material, problems known to arise in analysis and hypothesis for computational analysis and experimental techniques that were used in this study can be understood.

3.1 Anisotropic elasticity

3.1.1 LEFM for Anisotropic

We consider an elastic homogeneous anisotropic flat plate of uniform thickness which is in equilibrium as a result of the force distributed on its edge and the body forces. We assume that (1) at each point of the plate there is a plane of elastic symmetry which is parallel to the middle plane; (2) the force applied to the edge and the body force are acting within planes which are parallel to the middle plane and they are distributed symmetrically with respect to this plane and that they change slightly with the plate thickness; (3) deformations of the plate are small. A state of stress in a plate which satisfies the above conditions is called a state of generalized plane stress. Let x and y be a global Cartesian coordinate system. A local coordinate system x', y', z' is attached to the plane of transverse isotropy with the x'-axis taken normal to the plane and the y'-and z'-axis contained within the plane. A state of stress in a plate which satisfies the above conditions is called a state of generalized plane stress. The middle plane does not bend while undergoing deformation and remains plane. We take the middle plane as the coordinate plane x-y. Point 0 is the origin and x-axes and y-axes are directed arbitrarily (Figure 3.1).

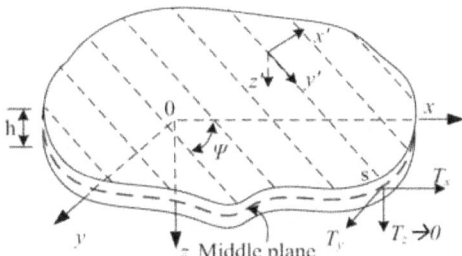

Figure 3.1: The definition of coordinate system on anisotropic plate.

The material orientation ψ is defined as the angle between the plane of transverse isotropy and the *x*-axis. We designate the plate thickness by h; *Tx* and *Ty*, the components of forces distributed at the edge per unit area; *X* and *Y*, the components of the body forces per unit volume (*Tz* = *Z* = 0 according to the conditions above). The constitutive relation of the body in the *x-y* plane is expressed as follows (Lekhnitskii, 1957)

$$\begin{Bmatrix} \varepsilon_x \\ \varepsilon_y \\ \gamma_{xy} \end{Bmatrix} = \begin{bmatrix} a_{11} & a_{12} & a_{16} \\ a_{21} & a_{22} & a_{26} \\ a_{61} & a_{62} & a_{66} \end{bmatrix} \begin{Bmatrix} \sigma_x \\ \sigma_y \\ \tau_{xy} \end{Bmatrix} \tag{3-1}$$

where a_{11}, a_{12}... a_{66} are the compliance components calculated in the *x-y* coordinate system. These compliance components depend on the elastic constants in the *x'*, *y'* coordinate system and the angle ψ. Using coordinate transformation rules, the compliance components in equation (3-1) are (Amadei, 1983)

$$a_{11} = \frac{sin^4 \psi}{E'} + \frac{cos^4 \psi}{E} + sin^2 \psi \, cos^2 \psi \left(\frac{1}{G'} - \frac{2v'}{E'}\right) \ ;$$

$$a_{12} = \left(\frac{1}{E'} + \frac{1}{E} - \frac{1}{G'}\right) sin^2 \psi \, cos^2 \psi - \frac{cos^4 \psi}{E'} v' - \frac{sin^4 \psi}{E'} v' \ ;$$

$$a_{16} = -\left[2\left(\frac{cos^2 \psi}{E} - \frac{sin^2 \psi}{E'}\right) + \left(\frac{1}{G'} - \frac{2v'}{E'}\right)(sin^2 \psi - cos^2 \psi)\right](sin \psi \, cos \psi) \ ;$$

$$a_{22} = \frac{cos^4 \psi}{E'} + \frac{sin^4 \psi}{E} + sin^2 \psi \, cos^2 \psi \left(\frac{1}{G'} - \frac{2v'}{E'}\right) \ ;$$

$$a_{26} = -\left[2\left(\frac{sin^2 \psi}{E} - \frac{cos^2 \psi}{E'}\right) - \left(\frac{1}{G'} - \frac{2v'}{E'}\right)(sin^2 \psi - cos^2 \psi)\right](sin \psi \, cos \psi) \ ;$$

33

$$a_{66} = \left(\frac{1}{E'} + \frac{1}{E} - \frac{2v'}{E'}\right) 4\sin^2\psi\cos^2\psi + \frac{1}{G'}(\sin^2\psi - \cos^2\psi)^2 \quad , \tag{3-2}$$

where E and E' denote the Young's modulus in a transversely isotropic plane and in a direction normal to the other, respectively; v and v' denote the Poisson's ratio characterizing the lateral strain response in a transversely isotropic plane to a stress acting parallel and in a direction normal to the other, respectively; G' denotes the shear modulus in a transversely isotropic plane. The shear modulus G along the plane of transverse isotropy is not independent and is equal to $E/(2(1+v))$.

The equilibrium equations are satisfied identically by introducing Airy' stress function F defined by (Timoshenko and Goodier, 1970)

$$\sigma_x = \frac{\partial^2 F}{\partial y^2}; \ \sigma_y = \frac{\partial^2 F}{\partial x^2}; \ \tau_{xy} = \frac{\partial^2 F}{\partial x \partial y} \tag{3-3}$$

the compatibility equation is given by

$$\frac{\partial^2 \varepsilon_x}{\partial y^2} + \frac{\partial^2 \varepsilon_y}{\partial x^2} = \frac{\partial^2 \gamma_{xy}}{\partial x \partial y} \tag{3-4}$$

substituting equation (3-3) into equation (3-1), and then into equation (3-4) we get the results in the following differential equation,

$$a_{22}\frac{\partial^4 F}{\partial x^4} - 2a_{26}\frac{\partial^4 F}{\partial x^3 \partial y} + (2a_{12} + a_{66})\frac{\partial^4 F}{\partial x^2 \partial y^2} - 2a_{16}\frac{\partial^4 F}{\partial x \partial y^3} + a_{11}\frac{\partial^4 F}{\partial y^4} = 0 \tag{3-5}$$

the general solution of this equation depends on the roots, μ_i (i=1 to 4), of its characteristic equation,

$$a_{11}\mu^4 - 2a_{16}\mu^3 + (2a_{12} + a_{66})\mu^2 - 2a_{26}\mu + a_{22} = 0 \tag{3-6}$$

Lekhnitskii (1963) has shown that the roots, two pairs of conjugate roots, of equation (3-6) are always either complex or purely imaginary. Let μ_1, μ_2 be the roots, which are assumed to be distinct, and $\overline{\mu_1}$, $\overline{\mu_2}$ be their respective conjugates. Substituting equation (3-2) into (3-6), it can shown that for a transversely isotropic plate and for a given inclination angle ψ, the roots depend on E/E', E/G' and v'. As shown by Lekhnitskii, the first derivatives of F with respect to x and y can be expressed as

$$\frac{\partial F}{\partial x} = 2Re[\phi_1(z_1) + \phi_2(z_2)] \ ;$$

34

$$\frac{\partial F}{\partial y} = 2Re[\mu_1\phi_1(z_1) + \mu_2\phi_2(z_2)] \tag{3-7}$$

where $\phi_k(z_k)$ (k=1, 2) are analytic functions of the complex variables $z_k = x + \mu_k y$, and $\textbf{\textit{Re}}$ denotes the real part of the complex expression in the brackets. Combining equation (3-3) and equation (3-7), we obtain the general expression for the stress components

$$\sigma_x = 2Re[\mu_1^2\phi_1'(z_1) + \mu_2^2\phi_2'(z_2)] \ ;$$
$$\sigma_y = 2Re[\phi_1'(z_1) + \phi_2'(z_2)] \ ; \tag{3-8}$$
$$\tau_{xy} = -2Re[\mu_1\phi_1'(z_1) + \mu_2\phi_2'(z_2)]$$

where $\phi_k'(z_k)$ are the first derivatives of $\phi_k(z_k)$ with respect to z_k. Substituting equation (3-8) into the constitutive relation and compatibility equation, the displacement components in the x and y directions are (Sih et al., 1965)

$$u = 2Re[A_{11}\phi_1(z_1) + A_{12}\phi_2(z_2)] \ ;$$
$$v = 2Re[A_{21}\phi_1(z_1) + A_{22}\phi_2(z_2)] \tag{3-9}$$

where the elements of the complex matrices A are as follows

$$A_{1j} = a_{11}\mu_j^2 + a_{22} - a_{16}\mu_j \ ;$$
$$A_{2j} = a_{12}\mu_j + \frac{a_{22}}{\mu_j} - a_{26} \ , \quad (j = 1, 2) \tag{3-10}$$

Furthermore, the tractions (as shown in Figure 3.1) components in the x and y directions are (Lekhnitskii, 1963)

$$T_x = 2Re[\mu_1\phi_1(z_1) + \mu_2\phi_2(z_2)] \ ;$$
$$T_y = -2Re[\phi_1(z_1) + \phi_2(z_2)] \tag{3-11}$$

here, the complex analytical functions $\phi_i(z_i)$ can in general express equations (3-8), (3-9), and equation (3-11) as follows (Suo, 1990, Lekhnitskii, 1963, and Ting, 1996)

$$u_i = 2Re\left[\sum_{j=1}^2 A_{ij}\phi_j(z_j)\right]; \quad T_i = -2Re\left[\sum_{j=1}^2 B_{ij}\phi_j(z_j)\right];$$
$$\sigma_{2i} = 2Re\left[\sum_{j=1}^2 B_{ij}\phi_j'(z_j)\right]; \quad \sigma_{1i} = -2Re\left[\sum_{j=1}^2 B_{ij}\phi_j'(z_j)\right], \ (i = 1, 2) \tag{3-12}$$

where $z_j = x+\mu_j y$; $\textbf{\textit{Re}}$ denotes the real part of a complex variable or function; a prime denotes the derivative; the complex number μ_j (j=1, 2) ; the elements of the complex matrices A is defined in equation (3-10) and the elements of the complex matrix $\boldsymbol{B}$ which can be defined as

$$B_{ij} = \begin{bmatrix} -\mu_1 & -\mu_2 \\ 1 & 1 \end{bmatrix} \tag{3-13}$$

3.1.2 Fundamental Solution (Green's function)

Assume that the medium is composed of two joined dissimilar anisotropic and elastic half-planes. We let the interface be along the x-axis, and the upper ($y > 0$) and lower ($y < 0$) half-planes be occupied by material (1) and (2), respectively (as shown in Figure 3.2).

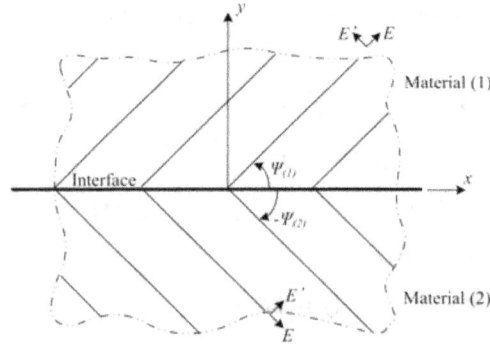

Figure 3.2: Definition of the coordinate systems within an anisotropic bi-material.

Considering a concentrated force acting at the source point (x^0, y^0) in material (2) ($y^0 < 0$), we express the complex vector function as (Suo, 1990)

$$\phi(z) = \begin{cases} \phi^U(z), & z \in (1) \\ \phi^L(z) + \phi^0_{(2)}(z), & z \in (2) \end{cases} \tag{3-14}$$

where the vector function is

$$\phi(z) = [\phi_1(z), \phi_2(z)]^T \tag{3-15}$$

with the argument having the generic form $z = x + \mu y$.

In equation (3-14), $\phi^0_{(2)}$ is a singular solution corresponding to a point force acting at the point (x^0, y^0) in an anisotropic infinite plane with the elastic properties of material (2). This singular solution can be expressed as (Suo, 1990 and Ting, 1996)

$$\phi_k^0(z_k) = -\frac{1}{2\pi} [H_{k1} P_1 ln(z_k - s_k) + H_{k2} P_2 ln(z_k - s_k)]$$ (3-16)

where $s_k = x^0 + \mu_k y^0$, and P_k is the magnitude of the point force in the k-direction, and

$$H = A^{-1} \left(Y^{-1} + \overline{Y}^{-1} \right)^{-1}; Y = iAB^{-1}$$ (3-17)

where $i = \sqrt{-1}$, overbar means the complex conjugate, superscript -1 means matrix inverse. There are two unknown vector functions, $\phi^U(z)$ and $\phi^L(z)$ solved in equation (3-14). While the former is analytic in the upper half plane (material (1)), the latter is analytic in the lower half-plane (material (2)). These expressions can be found by requiring continuity of the resultant traction and displacement across the interface, along with the standard analytic continuation arguments. Substituting equation (3-16) into equation (3-14), we got equation (3-18) as

$$\begin{cases} \phi^U(Z) = B_{(1)}^{-1} (Y_{(1)} + \overline{Y}_{(2)})^{-1} (\overline{Y}_{(2)} + Y_{(2)}) B_{(2)} \phi_{(2)}^0(z) & z \in (1) \\ \phi^L(Z) = B_{(2)}^{-1} (\overline{Y}_{(1)} + Y_{(2)})^{-1} (\overline{Y}_{(2)} - \overline{Y}_{(1)}) \overline{B}_{(2)} \phi_{(2)}^0(z) & z \in (2) \end{cases}$$ (3-18)

therefore, complex vector functions can show as

$$\phi(z) = \begin{cases} B_{(1)}^{-1} (Y_{(1)} + \overline{Y}_{(2)})^{-1} (\overline{Y}_{(2)} + Y_{(2)}) B_{(2)} \phi_{(2)}^0(z) & z \in (1) \\ B_{(2)}^{-1} (\overline{Y}_{(1)} + Y_{(2)})^{-1} (\overline{Y}_{(2)} - \overline{Y}_{(1)}) \overline{B}_{(2)} \phi_{(2)}^0(z) + \phi_{(2)}^0(z) & z \in (2) \end{cases}$$

(3-19)

In equation (3-19), the special subscript (1) and (2) are used exclusively to denote that the corresponding matrix or vector is in material (1) (y > 0) and material (2) (y < 0), respectively.

Similarly, for a point force in material (1) ($y^0 > 0$), these complex vector functions can be found as

$$\phi(z) = \begin{cases} B_{(1)}^{-1}(\overline{Y}_{(2)} + Y_{(1)})^{-1}(\overline{Y}_{(1)} - \overline{Y}_{(2)})\overline{B}_{(1)}\overline{\phi}_{(1)}^{0}(z) + \phi_{(1)}^{0}(z) & z \in (1) \\ B_{(2)}^{-1}(Y_{(2)} + \overline{Y}_{(1)})^{-1}(\overline{Y}_{(1)} + Y_{(1)})B_{(1)}\phi_{(1)}^{0}(z) & z \in (2) \end{cases}$$

$$(3\text{-}20)$$

where the vector functions $\phi_{(1)}^{0}$ is the infinite-plane solution given in equation (3-16) but with the elastic properties of material (1).

With the complex vector function given in equation (3-19) and equation (3-20), the Green's functions of the displacement and traction can be obtained by substituting these complex functions into equation (3-12). Here, U_{kl}^{*} is the Green's function for displacement and T_{kl}^{*} is the Green's function for traction. Special superscripts and subscripts (1) and (2) are used exclusively to denote that the corresponding quantities are on materials (1) (y > 0) and (2) (y < 0), respectively. (Pan and Amadei, 1999)

(i). For source point (s) and field point (z) on material (1) (y > 0)

$$U_{kl}^{*} = \frac{-1}{\pi} Re \left\{ \sum_{j=1}^{2} A_{lj}^{(1)} \left[ln\left(z_{j}^{(1)} - s_{j}^{(1)}\right) H_{jk}^{(1)} + \sum_{i=1}^{2} W_{ji}^{11} ln\left(z_{j}^{(1)} - s_{i}^{(1)}\right) \overline{H}_{ik}^{(1)} \right] \right\}$$

$$(3\text{-}21)$$

$$T_{kl}^{*} = \frac{1}{\pi} Re \left\{ \sum_{j=1}^{2} B_{lj}^{(1)} \left[\frac{\mu_{j}^{(1)} n_x - n_y}{z_{j}^{(1)} - s_{j}^{(1)}} H_{jk}^{(1)} + \sum_{i=1}^{2} W_{ji}^{11} \frac{\mu_{j}^{(1)} n_x - n_y}{z_{j}^{(1)} - s_{i}^{(1)}} \overline{H}_{ik}^{(1)} \right] \right\} \qquad (3\text{-}22)$$

where the matrix H is defined in equation (3-17) with the anisotropic elastic properties of material (1), and

$$W^{11} = B_{(1)}^{-1}(Y_{(1)} + \overline{Y}_{(2)})^{-1}(\overline{Y}_{(1)} - \overline{Y}_{(2)})\overline{B}_{(1)}$$

$$(3\text{-}23)$$

(ii).For source point (s) and field point (z) on material (1) (y > 0), and field point (z) on material (2) (y < 0)

$$U_{kl}^{*} = \frac{-1}{\pi} Re \left\{ \sum_{j=1}^{2} A_{lj}^{(2)} \left[\sum_{i=1}^{2} W_{ji}^{12} ln\left(z_{j}^{(2)} - s_{i}^{(1)}\right) H_{ik}^{(1)} \right] \right\}$$

$$(3\text{-}24)$$

$$T_{kl}^{*} = \frac{1}{\pi} Re \left\{ \sum_{j=1}^{2} B_{lj}^{(2)} \left[\sum_{i=1}^{2} W_{ji}^{12} \frac{\mu_{j}^{(2)} n_x - n_y}{z_{j}^{(2)} - s_{i}^{(1)}} \overline{H}_{ik}^{(1)} \right] \right\}$$

$$(3\text{-}25)$$

with

$$W^{12} = B_{(2)}^{-1}(Y_{(2)} + \overline{Y}_{(1)})^{-1}(\overline{Y}_{(1)} - Y_{(1)})B_{(1)} \tag{3-26}$$

(iii). For source point (s) and field point (z) on material (2) ($y < 0$)

$$U_{kl}^* = \frac{-1}{\pi} Re\left\{\sum_{j=1}^{2} A_{lj}^{(2)}\left[ln(z_j^{(2)} - s_j^{(2)})H_{jk}^{(2)} + \sum_{i=1}^{2} W_{ji}^{22}ln(z_j^{(2)} - s_i^{(2)})\overline{H}_{ik}^{(2)}\right]\right\} \tag{3-27}$$

$$T_{kl}^* = \frac{1}{\pi} Re\left\{\sum_{j=1}^{2} B_{lj}^{(2)}\left[\frac{\mu_j^{(2)}n_x - n_y}{z_j^{(2)} - s_j^{(2)}}H_{jk}^{(2)} + \sum_{i=1}^{2} W_{ji}^{22}\frac{\mu_j^{(2)}n_x - n_y}{z_j^{(2)} - s_i^{(2)}}\overline{H}_{ik}^{(2)}\right]\right\} \tag{3-28}$$

where the matrix H is defined in equation (3-17) with the anisotropic elastic properties of material (2), and

$$W^{22} = B_{(2)}^{-1}(Y_{(2)} + \overline{Y}_{(1)})^{-1}(\overline{Y}_{(2)} - \overline{Y}_{(1)})\overline{B}_{(2)} \tag{3-29}$$

(iv). For source point (s) and field point (z) on materials (2) ($y < 0$), and field point (z) on material 1 ($y > 0$)

$$U_{kl}^* = \frac{-1}{\pi} Re\left\{\sum_{j=1}^{2} A_{lj}^{(1)}\left[\sum_{i=1}^{2} W_{ji}^{21}ln(z_j^{(1)} - s_i^{(2)}) H_{ik}^{(2)}\right]\right\} \tag{3-30}$$

$$T_{kl}^* = \frac{1}{\pi} Re\left\{\sum_{j=1}^{2} B_{lj}^{(1)}\left[\sum_{i=1}^{2} W_{ji}^{21}\frac{\mu_j^{(1)}n_x - n_y}{z_j^{(1)} - s_i^{(2)}}\overline{H}_{ik}^{(2)}\right]\right\} \tag{3-31}$$

with

$$W^{21} = B_{(1)}^{-1}(Y_{(1)} + \overline{Y}_{(2)})^{-1}(\overline{Y}_{(2)} - Y_{(2)})B_{(2)} \tag{3-32}$$

It is noteworthy that these Green's function can be used to solve both plane stress and plane strain problems on anisotropic bi-materials. Although the isotropic solution cannot be analytically reduced from these Green's functions, one can numerically approximate it by selecting a very weak anisotropic (or nearly isotropic) medium (Pan and Amadei, 1996, Sollero et al., 1994).

3.2 Boundary Element Method formulation for 2D Cracked Anisotropic Bi-materials

The traditional displacement boundary integral equation for linear elasticity can be expressed as

$$b_{ij}(s_k^0)u_j(s_k^0) + \int_\Gamma T_{ij}^*(z_k, s_k^0)u_j(z_k)\,d\Gamma(z_k) = \int_\Gamma U_{ij}^*(z_k, s_k^0)t_j(z_k)\,d\Gamma(z_k)$$

$$(3\text{-}33)$$

where $i, j, k = 1, 2$; T_{ij} and U_{ij} are the Green's tractions and displacement given in equations (3-12) and (3-13); u_j and t_j are the boundary displacements and tractions; b_{ij} are quantities that depend on the geometry of the boundary and are equal to $\delta_{ij}/2$ for a smooth boundary; z_k and s_k^0 are the field and source points on the boundary Γ of the domain. Discretization of equation (3-33) gives a linear system of algebraic equations, which can be solved for the unknown displacements u_j and tractions t_j on the boundary. However, for a cracked elastic medium, equation (3-33) is not sufficient for solving all the unknowns along the outer boundary of the problem as well as along two sides of the crack surfaces because of the geometric singularity associated with the crack surface.

It is well known that a cracked domain poses certain difficulties for BEM modeling (Cruse, 1988). Previously, fracture mechanics problems in isotropic or anisotropic bimaterials were mostly handled by the multi-domain method in which each side of the crack surface is put into different domains and artificial boundaries are introduced to connect the crack surface to the un-cracked boundary. For the bimaterial case, discretization along the interface is also required if one uses the Kelvin-type (infinite domain) Green's functions.

In this section, we present a single-domain BEM formulation in which neither the artificial boundary nor the discretization along the un-cracked interface is necessary. This single-domain BEM formulation was widely used recently by

Chen et al., 1996 for homogeneous materials and is now extended to anisotropic bi-materials.

For a point $s_{k,B}^0$ on the un-cracked boundary, the displacement integral equation applied to the outer boundary results in the following form ($s_{k,B}^0 \in \Gamma_B$ only, as shown in Figure 3.3)

$$b_{ij}\left(s_{k,B}^0\right)u_j\left(s_{k,B}^0\right) + \int_{\Gamma_B} T_{ij}^*\left(z_{k,B}, s_{k,B}^0\right) u_j\left(z_{k,B}\right) d\Gamma\left(z_{k,B}\right)$$

$$+ \int_{\Gamma_C} T_{ij}^*\left(z_{k,C}, s_{k,B}^0\right) \left[u_j\left(z_{k,C+}\right) - u_j\left(z_{k,C-}\right)\right] d\Gamma\left(z_{k,C}\right) \qquad (3\text{-}34)$$

$$= \int_{\Gamma_B} U_{ij}^*\left(z_{k,B}, s_{k,B}^0\right) t_j\left(z_{k,B}\right) d\Gamma\left(z_{k,B}\right)$$

where $i, j, k=1, 2$; T_{ij}^* and U_{ij}^* are Green's tractions and displacements given in equations (3-21), (3-22), (3-24), (3-25), (3-27), (3-28), (3-30), and equation (3-31); u_j and t_j are the boundary displacement and tractions, respectively; z_k and $s_{k,B}^0$ are the field points and the source points on the boundary Γ of the domain. Γ_C has the same outward normal as Γ_{C+}. b_{ij} are coefficients that depend only upon the local geometry of the un-cracked boundary at point $s_{k,B}^0$ and are equal to $\delta_{ij}/2$ for a smooth boundary; Here, the subscripts B and C denote the outer boundary and the crack surface, respectively. In deriving equation (3-34), we have assumed that the tractions on the two faces of a crack are equal and opposite. We emphasize here that since the bi-material Green's functions are included in equation (3-34), discretization along the interface can be avoided, with the exception of the interfacial crack part which will be treated by the traction integral equation presented below.

It is noted that all the terms on the right-hand side of equation (3-34) have only weak singularities, thus, are integrable. Although the second term on the left-hand side of equation (3-34) has a strong singularity, it can be treated by the rigid-body motion method.

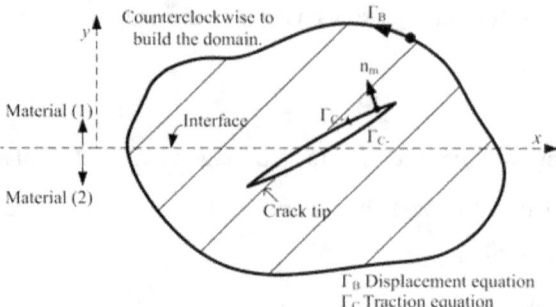

Figure 3.3: Geometry of a 2-D cracked domain.

The traction integral equation (for s_k^0 being a smooth point on the crack) applied to one side of the crack surfaces is ($s_{k,C}^0 \in \Gamma_{C+}$ only)

$$0.5 t_j\left(s_{k,C}^0\right) + n_m\left(s_{k,C}^0\right) \int_{\Gamma_B} C_{limk} T_{ij,k}^*\left(s_{k,C}^0, z_{k,B}\right) u_j\left(z_{k,B}\right) d\Gamma\left(z_{k,B}\right)$$

$$+ n_m\left(s_{k,C}^0\right) \int_{\Gamma_C} C_{limk} T_{ij,k}^*\left(s_{k,C}^0, z_{k,C}\right)\left[u_j\left(z_{k,C+}\right) - u_j\left(z_{k,C-}\right)\right] d\Gamma\left(z_{k,C}\right) \qquad (3\text{-}35)$$

$$= n_m\left(s_{k,C}^0\right) \int_{\Gamma_B} C_{limk} U_{ij,k}^*\left(s_{k,C}^0, z_{k,B}\right) t_j\left(z_{k,B}\right) d\Gamma\left(z_{k,B}\right)$$

where n_m is the unit outward normal at the crack surface $s_{k,C}^0$ and C_{limk} is the fourth-order stiffness tensor.

Equation (3-34) and equation (3-35) form a pair of boundary integral equation (Chen, 1999, Pan, 1997, and Crouch, 1983) and can be used to calculate SIFs in anisotropic bi-materials. The main feature of the BEM formulation is that it is a single-domain formulation with the displacement integral equation (3-34) being collocated on the un-cracked boundary only and the traction integral equation (3-35) on one side of the crack surface only. For problems without cracks, one needs equation (3-34) only, with the integral on the crack surface being discarded. Equation (3-34) then reduces to the well-known displacement integral on the un-cracked boundary.

The internal stress $\sigma(z_k)$ are determined by the following expression.

$$\sigma_{lm}(z_k) + \int_{\Gamma_B} C_{limk} T_{ij,k}^*\left(s_{k,C}^0, z_{k,B}\right) u_j\left(z_{k,B}\right) d\Gamma\left(z_{k,B}\right)$$

$$+ \int_{\Gamma C} C_{limk} T^*_{ij,k}\left(s^0_{k,C}, z_{k,C}\right)\left[u_j\left(z_{k,C+}\right) - u_j\left(z_{k,C-}\right)\right] d\Gamma\left(z_{k,C}\right) \qquad (3\text{-}36)$$

$$= \int_{\Gamma B} C_{limk} U^*_{ij,k}\left(s^0_{k,C}, z_{k,B}\right) t_j\left(z_{k,B}\right) d\Gamma\left(z_{k,B}\right)$$

For given particular solutions (related to the body force of gravity, rotational forces, and the far-field stresses), the boundary integral equations (3-34) and (3-35) can be discretized and solved numerically for the unknown boundary displacements (or displacement discontinuities on the crack surface) and tractions. In solving these equations, the hyper-singular integral term involved in equation (3-35) can be handled by an accurate and efficient Gauss quadrature formulae which is similar to the traditional weighted Gauss quadrature but with a different weight (Tsamasphyros and Dimou, 1990).

3.3 Calculation of stress intensity factor

In fracture mechanics analysis, especially in the calculation of SIFs, one needs to know the asymptotic behavior of the displacements and stresses near the crack-tip. In our BEM analysis of SIFs, we propose to use the extrapolation method of the crack-tip displacements. We therefore need to know the exact asymptotic behavior of the relative crack displacement (RCD) behind the crack-tip. This asymptotic expression has different forms depending on the location of the crack-tip. In this study, two cases will be discussed, that is, a crack-tip within the homogeneous material and an interfacial crack-tip.

3.3.1 A crack tip with in a homogeneous material

Assume that the crack-tip is within material (1) , the asymptotic behavior of the relative displacement at a distance r behind the crack-tip can be expressed in terms of the three SIFs as (Suo, 1990)

$$\Delta u(r) = 2\sqrt{\frac{2r}{\pi}} \text{Re}\left(Y_{(1)}\right) K \qquad (3\text{-}37)$$

where $K = [K_{II}, K_I, K_{III}]^T$ is the SIF vector, and Y is a matrix with elements related to the anisotropic properties in material (1), as defined in equation (3-17).

In order to get the square-root characteristic of the relative crack displacement (RCD) near crack tip exactly, we construct the following new crack tip element with the tip at $\xi = -1$ (as shown in Figure 3.5(e))

$$\Delta u_i = \sum_{k=1}^{3} f_k \Delta u_i^k \qquad (3\text{-}38)$$

where the subscript i denotes RCD component and the superscript k (1,2,3) denotes the RCDs at nodes $\xi = -\frac{2}{3}, 0, \frac{2}{3}$, (as shown in Figure 3.5(b)) respectively. The shape functions f_k are those introduced by (Pan, 1997)

$$f_1 = \frac{3\sqrt{3}}{8}\sqrt{\xi + 1}[5 - 8(\xi + 1) + 3(\xi + 1)^2],$$

$$f_2 = \frac{1}{4}\sqrt{\xi + 1}[-5 + 18(\xi + 1) + 9(\xi + 1)^2] \qquad (3\text{-}39)$$

$$f_3 = \frac{3\sqrt{3}}{8\sqrt{5}}\sqrt{\xi + 1}[1 - 4(\xi + 1) + 3(\xi + 1)^2]$$

For this case, the relation of the RCDs at a distance r behind the crack-tip and the SIFs can be found as (Sih et al., 1965, Sollero and Aliabadi, 1993, and Pan, 1997)

$$\Delta u_1 = 2\sqrt{\frac{2r}{\pi}}(H_{11}K_I + H_{12}K_{II})$$
$$\Delta u_2 = 2\sqrt{\frac{2r}{\pi}}(H_{21}K_I + H_{22}K_{II}) \qquad (3\text{-}40)$$

where

$$H_{11} = Im\left(\frac{\mu_2 P_{11} - \mu_2 P_{12}}{\mu_1 - \mu_2}\right); \quad H_{12} = Im\left(\frac{P_{11} - P_{12}}{\mu_1 - \mu_2}\right)$$
$$H_{21} = Im\left(\frac{\mu_2 P_{21} - \mu_2 P_{22}}{\mu_1 - \mu_2}\right); \quad H_{22} = Im\left(\frac{P_{21} - P_{22}}{\mu_1 - \mu_2}\right) \qquad (3\text{-}41)$$

Substituting the RCDs into the equation (3-38) and equation (3-40), we obtain a set of algebraic equations which the SIFs K_I and K_{II} can be solved.

3.3.2 An interfacial crack tip

For this case, the relative crack displacements at a distance r behind the interfacial crack-tip can be expressed, in terms of the three SIFs, as (Gao et al., 1992)

$$\Delta u(r) = \left(\sum_{j=1}^{3} c_j D Q_j e^{-\pi \delta_j} r^{\frac{1}{2}+\delta_j} \right) K \tag{3-42}$$

where c_j, δ_j, Q_j and D are the relative parameters in material (1) and material (2). Utilizing equation (3-17), we defined the matrix of material as

$$Y_{(1)} + \overline{Y}_{(2)} = D - iV \tag{3-43}$$

where, D and V are two real matrices, and then utilizing these two matrices to define matrix P as

$$P = -D^{-1}V \tag{3-44}$$

and the characteristic β relative to material

$$\beta = \sqrt{-\frac{1}{2} tr(P^2)} \tag{3-45}$$

After, we used the characteristic β we obtained to defined oscillation index ε as

$$\varepsilon = \frac{1}{2\pi} ln \frac{1+\beta}{1-\beta} = \frac{1}{\pi} tanh^{-1} \beta$$
$$\delta_1 = 0, \delta_2 = \varepsilon, \delta_3 = -\varepsilon \tag{3-46}$$
$$Q_1 = P^2 + \beta^2 I, \quad Q_2 = P(P - i\beta I), \quad Q_3 = P(P + i\beta I)$$

where I is a 3×3 identity matrix.

Utilizing the relationship between the characteristic β and oscillation index ε to define constant c_j as

$$c_1 = \frac{2}{\sqrt{2\pi}\beta^2}; \quad c_2 = \frac{-e^{-\pi\varepsilon} d^{i\varepsilon}}{\sqrt{2\pi}(1+2i\varepsilon)\beta^2 \cosh(\pi\varepsilon)}; \quad c_3 = \frac{-e^{-\pi\varepsilon} d^{i\varepsilon}}{\sqrt{2\pi}(1-2i\varepsilon)\beta^2 \cosh(\pi\varepsilon)} \tag{3-47}$$

where d is the characteristic distance along material interface to crack tip.

Comparing equation (3-37) with equation (3-42), we noticed that while the relative crack displacement behaves as a square-root for a crack-tip within a homogeneous medium, for an interfacial crack-tip, its behavior is $r^{\frac{1}{2}+i\delta}$, a square-root feature multiplied by weak oscillatory behaviors.

Equation (3-42) can be recast into the following form, which is more convenient for the current numerical applications:

$$\Delta u(r) = \sqrt{\frac{2r}{\pi}} M\left(\frac{r}{d}\right) K \tag{3-48}$$

where d is the characteristic length, and M is a matrix function with its expression given by

$$M(x) = \frac{D}{\beta^2}\left\{(P^2 + \beta^2 I) - \frac{[cos(\varepsilon \ln x) + 2\varepsilon \sin(\varepsilon \ln x)]P^2 + \beta[sin(\varepsilon \ln x) - 2\varepsilon \cos(\varepsilon \ln x)]P}{(1+4\varepsilon^2)\cosh(\pi\varepsilon)}\right\} \tag{3-49}$$

Again, in order to capture the square-root and the weak oscillatory behavior, we construct a crack-tip element with tip at $\xi = -1$ (as shown in Figure 3.5(e)) in terms of which the relative crack displacement is expressed as

$$\Delta u(r) = M\left(\frac{r}{d}\right)\begin{bmatrix} f_1\Delta u_1^1 & f_2\Delta u_1^2 & f_3\Delta u_1^3 \\ f_1\Delta u_2^1 & f_2\Delta u_2^2 & f_3\Delta u_2^3 \\ f_1\Delta u_3^1 & f_2\Delta u_3^2 & f_3\Delta u_3^3 \end{bmatrix} \tag{3-50}$$

3.3.3 Construction of numerical model

In this study, the numerical model of two dimensional problems has been built by the three geometric nodes quadratic element. Following, we will introduce to the three geometric nodes quadratic elements which how to apply to the boundary and crack of 2D problem.

In general, the kind of element has been used to model two dimensional numerical problems can be divided in to (i) constant elements (figure 3.4(a)), (ii) linear elements (figure 3.4(b)), and (iii) curved elements such as the quadratic ones shown in figure 3.4(c). This figure 4.1 shows that the quadratic element is in slightly better condition than constant element and linear element on the two

46

dimensional irregular boundary problems. (Brebbia and Dominguez, 1992)

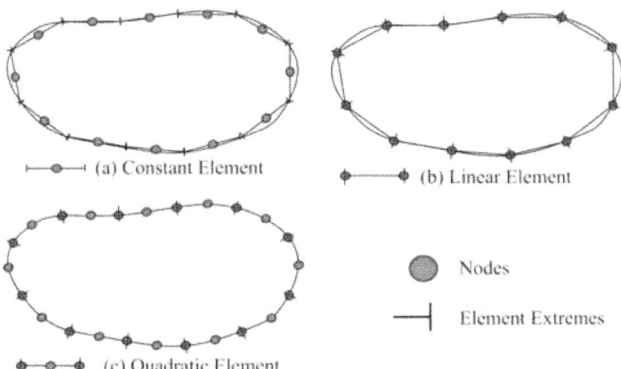

Figure 3.4: Different types of boundary elements

Quadratic elements are very often used to model problems with curvilinear geometry. In analyzing the three geometric nodes quadratic elements are subdivided in to five types as shown in Figure 3.5. For these elements, both the geometry and the boundary quantities are approximated by intrinsic coordinate ξ, and the shape functions f_k can be fined in equation (3-39). Generally the three geometric nodes of an element are used as collocation points. Such an element is referred to as a continuous quadratic element of *Type* V (as shown in Figure 3.5(d)). To model the corner points or the points where there is a change in the boundary conditions, one sided discontinuous quadratic elements (commonly referred to as the partially discontinuous elements) are used. These elements can be of four types depending on which extreme collocation point has been shifted inside the element to model the geometric or physical singularity. If it is the third collocation node which has been shifted inside, then the resulting element is referred to as the partially discontinuous quadratic element of *Type* I (as shown in Figure 3.5(a)). Otherwise, if the first node is shifted inside, then the element is called the partially discontinuous quadratic element of *Type* III (as shown in Figure 3.5(c)). One may opt for moving both extreme nodes inside which results in a discontinuous elements of *Type* II and *Type* VI, (to model the crack surface or the points where there is in the crack tip positions) such as shown in Figure

47

3.5(b), and Figure 3.5(e). In the graphical representation of these elements, we have used a "×" for a geometric node, a "○" for a collocation node, and a "◊" for a crack tip position.

In meshing the 2D anisotropic bi-material problem (as shown in Figure 3.6), we assume the interface be along the x-axis, and the upper ($y > 0$) and lower ($y < 0$) half-planes be occupied by material (1) and (2), respectively. The corner of outer boundary is processed by the discontinuous elements *Type* I and *Type* III; the continuous element *Type* V to deal with all outer smooth boundary; internal crack surface is processed by crack surface elements *Type* II; and crack tip element *Type* VI is process to crack tip problem. In order to avoid the oscillatory behavior of the interface, we mesh an anisotropic problem (as shown in Figure 3.6) of bi-material in which neither the interfacial elements nor the discretization along the un-cracked interface is necessary, with the exception of the interfacial crack part which will be treated by the traction integral equation (3-35) and equation (3-42) of RCD.

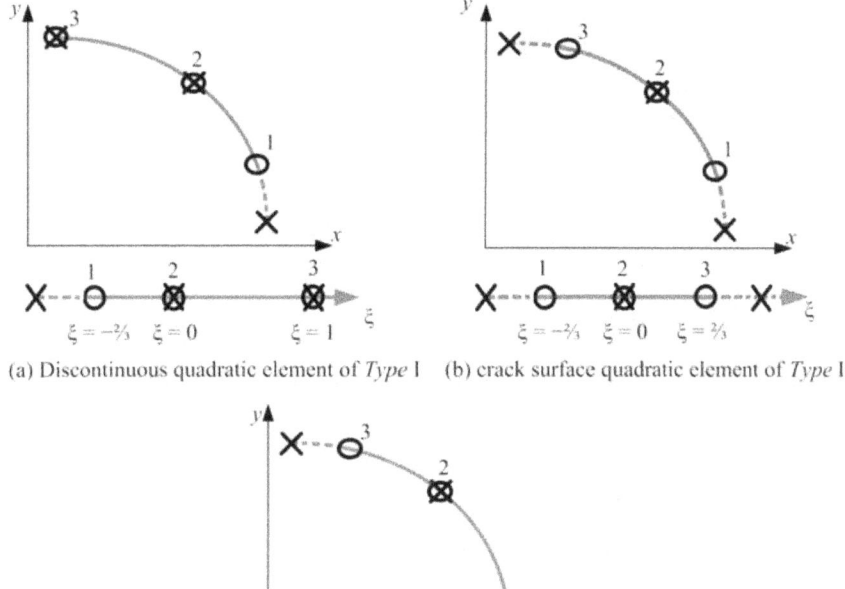

(a) Discontinuous quadratic element of *Type* I   (b) crack surface quadratic element of *Type* II

(c) Discontinuous quadratic element of *Type* III

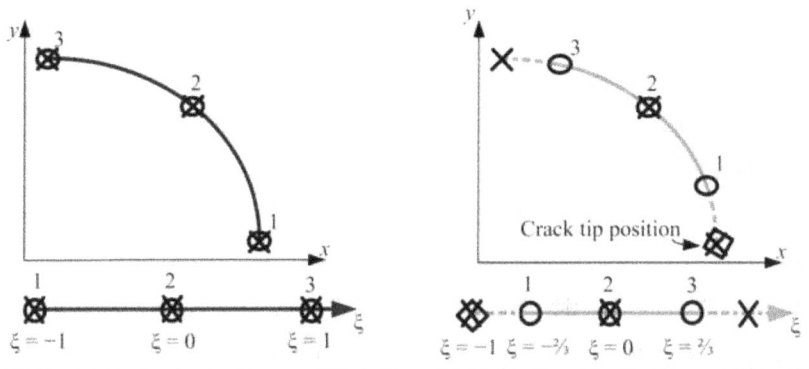

(d) Continuous quadratic element of *Type* V   (e) Crack tip quadratic element of *Type* VI

Figure 3.5: The three geometric node quadratic elements are used to approximate the un-crack boundary and crack for two dimensional problems.

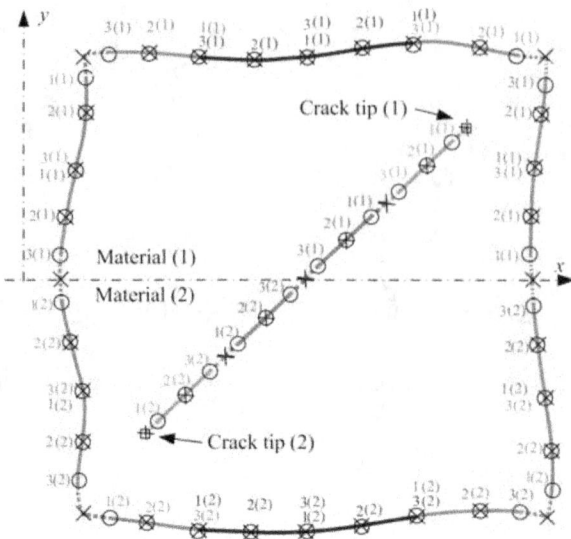

Figure 3.6: Mesh 2D bi-material problem with the five types of quadratic
elements

3.4 Crack initiation and fracture propagation

In fracture mechanics, there are three criteria commonly used to predict the crack initiation angle: the maximum tangential stress criterion, or σ-criterion (Erdogan and Sih, 1963); the maximum energy release rate criterion, or G-criterion (Palaniwamy and Knauss, 1972) and the minimum strain energy density criterion, or S-criterion (Sih, 1974). Among them, the σ-criterion has been found to predict well the directions of crack initiation compared to the experimental results for polymethylmethacrylate (Woo and Ling, 1984, and Richard, 1984) and brittle clay (Vallejo, 1987). Because of its simplicity, the σ-criterion seems to be the most popular criterion in mixed mode I-II fracture studies (Whittaker et al., 1992). Therefore, the σ-criterion was used in this paper to determine the crack initiation angle for anisotropic plates to predict the crack propagation.

3.4.1 Crack initiation angle

For anisotropic materials, the general form of the elastic stress field near the crack tip in the local Cartesian coordinates $x'' - y''$ in Figure 3.7 can be expressed in terms of the two stress intensity factors K_I and K_{II} as follows (Sih et al., 1965)

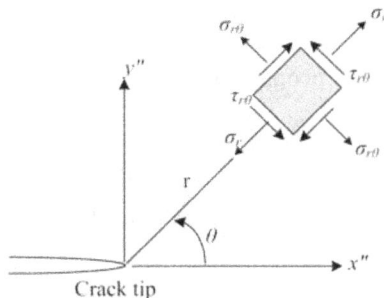

Figure 3.7: Crack tip coordinate system and stress components.

$$\sigma_{x''} = \frac{K_I}{\sqrt{2\pi r}} Re\left[\frac{\mu_1\mu_2}{\mu_1-\mu_2}\left(\frac{\mu_2}{\sqrt{\cos\theta+\mu_2\sin\theta}} - \frac{\mu_1}{\sqrt{\cos\theta+\mu_1\sin\theta}}\right)\right]$$
$$+ \frac{K_{II}}{\sqrt{2\pi r}} Re\left[\frac{1}{\mu_1-\mu_2}\left(\frac{\mu_2^2}{\sqrt{\cos\theta+\mu_2\sin\theta}} - \frac{\mu_1^2}{\sqrt{\cos\theta+\mu_1\sin\theta}}\right)\right] \qquad (3\text{-}51)$$

$$\sigma_{y''} = \frac{K_I}{\sqrt{2\pi r}} Re\left[\frac{1}{\mu_1-\mu_2}\left(\frac{\mu_1}{\sqrt{\cos\theta+\mu_2\sin\theta}} - \frac{\mu_2}{\sqrt{\cos\theta+\mu_1\sin\theta}}\right)\right]$$
$$+ \frac{K_{II}}{\sqrt{2\pi r}} Re\left[\frac{1}{\mu_1-\mu_2}\left(\frac{1}{\sqrt{\cos\theta+\mu_2\sin\theta}} - \frac{1}{\sqrt{\cos\theta+\mu_1\sin\theta}}\right)\right] \qquad (3\text{-}52)$$

$$\tau_{x''y''} = \frac{K_I}{\sqrt{2\pi r}} Re\left[\frac{\mu_1\mu_2}{\mu_1-\mu_2}\left(\frac{1}{\sqrt{\cos\theta+\mu_1\sin\theta}} - \frac{1}{\sqrt{\cos\theta+\mu_2\sin\theta}}\right)\right]$$
$$+ \frac{K_{II}}{\sqrt{2\pi r}} Re\left[\frac{1}{\mu_1-\mu_2}\left(\frac{\mu_1}{\sqrt{\cos\theta+\mu_1\sin\theta}} - \frac{\mu_2}{\sqrt{\cos\theta+\mu_2\sin\theta}}\right)\right] \qquad (3\text{-}53)$$

Using coordinate transformation, the stress fields near the crack tip in the polar coordinates (r, θ) in Figure 3.4 is as follows

$$\sigma_\theta = \frac{\sigma_{x''}+\sigma_{y''}}{2} - \frac{\sigma_{x''}-\sigma_{y''}}{2}\cos 2\theta - \tau_{x''y''}\sin 2\theta$$
$$\tau_\theta = -\frac{\sigma_{x''}-\sigma_{y''}}{2}\sin 2\theta + \tau_{x''y''}\cos 2\theta \qquad (3\text{-}54)$$

If the maximum σ-criterion is used, the angle of crack initiation, θ_0, must satisfy

$$\frac{\partial \sigma_\theta}{\partial \theta} = 0 \ (or \ \tau_{r\theta} = 0) \ and \ \frac{\partial^2 \sigma_\theta}{\partial \theta^2} < 0 \tag{3-55}$$

A numerical procedure was applied to find the angle θ_0 when σ_θ is a maximum for known values of the material elastic constants, the anisotropic orientation angle ψ and the crack geometry.

3.4.2 Simulation of crack propagation path steps

In this paper, the process of crack propagation in an anisotropic homogeneous plate under mixed mode I-II loading is simulated by incremental crack extension with a piece-wise linear discretization. For each incremental analysis, crack extension is conveniently modeled by a new boundary element. A computer program has been developed to automatically generate new data required for analyzing sequentially the changing boundary configuration. Based on the calculation of the SIFs and crack initiation angle for each increment, the procedure of crack propagation can be simulated. The steps in the crack propagation process are summarized as follows (Figure 3.8):

(i)   Compute SIFs by using the proposed BEM;

(ii)   Determine an angle of crack initiation based on the maximum tensile stress criterion;

(iii) Extend a new crack by a linear element along the direction determined in step (ii) (We will discuss that how select the length of linear element to mesh new extension crack in the following chapter.);

(iv)   Automatically generate new BEM meshes;

(v)   Repeat all the above steps until crack is near the outer boundary.

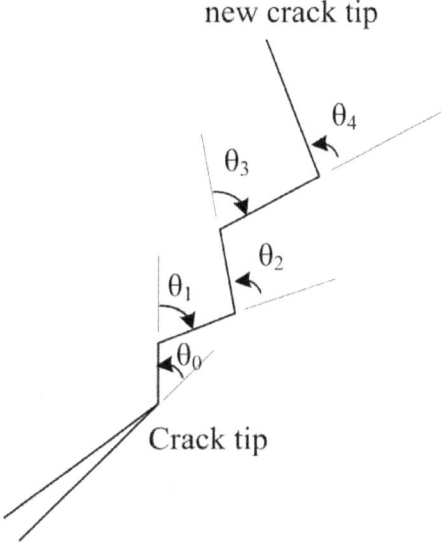

Figure 3.8: Process of crack propagation by increasing the number of
linear elements

Chapter 4

Verification of the Program

Based on the theory background of chapter 3, the computer code, developed by formula translation (FORTRAN) code, can be used to calculate the crack initiation angle (θ_0), propagation path, and stress intensity factors (SIFs, K_I and K_{II}) with different crack angles (β), crack length ($2a$) and anisotropic orientation (ψ) of an anisotropic bi-material. In this chapter, we show the numerical results that have been verified by analytical solution of literature to establish correctness of the computer program.

4.1 Determination the number of quadratic elements for two dimensional bi-material problems

The accuracy of numerical results is affected by the number of elements for crack surface or outer boundary. Moreover, the number of elements is profoundly dependent by subjective judgment of the analyst. Therefore, how determinate the number of elements which is meshing on crack surface (or outer boundary) to accurate calculate SIFs is very important. The number of elements for crack surface has been judged through an analysis of infinite domain problem; and the number of elements for outer boundary has been judged from an analysis of finite domain problem, such as follows.

4.1.1 Determination of the number of element for crack surface

Consider a vertical crack intersecting an interface and subjected to far-field horizontal stresses as shown in Figure 4.1 in an infinite domain of isotropic bi-material. The horizontal far field stresses applied in materials (1) and (2) are, respectively, $\sigma_{(1)}$ and $\sigma_{(2)}(= \sigma_{(1)} G_{(2)}/G_{(1)})$, the special subscript (1) and (2) are in materials (1) and (2), respectively. The Poisson ratio $v_{(1)}$ and $v_{(2)}$ are assumed to be equal to 0.3 and the shear modulus ratio $G_{(2)}/G_{(1)}$ is assumed to be equal to 0.3. The distance of the crack tip A and B to the interface are equal to a, the half-length of the crack. A plane stress condition is assumed.

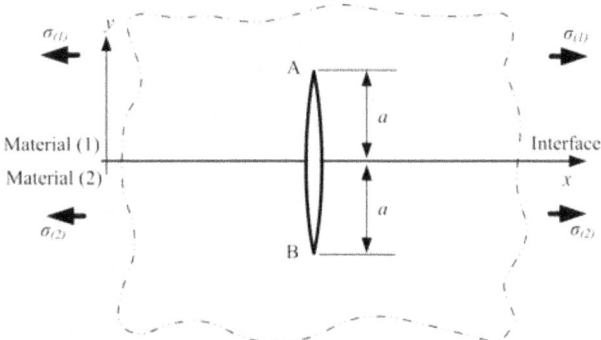

Figure 4.1: A vertical crack intersecting an interface under far-field stresses.

In order to determinate the number of elements for crack surface, Hence the number of quadratic elements were used to discretize the crack surface (a) is to vary. The SIFs at the crack tips A and B are listed in Table 4.1 for several values of the number of elements, and are compared with the analytical solutions obtained by Isida and Noguchi, 1983. As shown in Figure 4.2, the curve of crack tips A and B becomes gentle after 10 quadratic elements; hence we can find that the results between analytical and numerical approaches are very close when crack surface with the number of quadratic elements is equal to 10. It is obvious that 10 or more discontinuous quadratic elements are suitable for determining the SIFs.

Table 4.1 Calculation for SIFs with several values of the number of elements for crack surface.

Number of elements	$K_I/\sigma\sqrt{\pi a}$ of tip A			$K_I/\sigma\sqrt{\pi a}$ of tip B		
	Present	Isida et al. (1983)	Diff. (%)	Present	Isida et al. (1983)	Diff. (%)
2	1.016122		-0.11	1.063321		0.06
4	1.015938		-0.09	1.063636		0.03
6	1.015821		-0.08	1.063799		0.02
8	1.015749		-0.07	1.063869		0.01
10	1.015711		-0.07	1.063900		0.01
12	1.015692		-0.07	1.063916		0.01
14	1.015682		-0.07	1.063933		0.01
16	1.015677		-0.07	1.063933		0.01
18	1.015675	1.015	-0.07	1.063945	1.064	0.01
20	1.015672		-0.07	1.063945		0.01
22	1.015667		-0.07	1.063951		0.00
24	1.015668		-0.07	1.063956		0.00
26	1.015660		-0.07	1.063956		0.00
28	1.015655		-0.06	1.063956		0.00
30	1.015645		-0.06	1.063951		0.00
32	1.015639		-0.06	1.063956		0.00
34	1.015636		-0.06	1.063956		0.00
36	1.015634		-0.06	1.063951		0.00

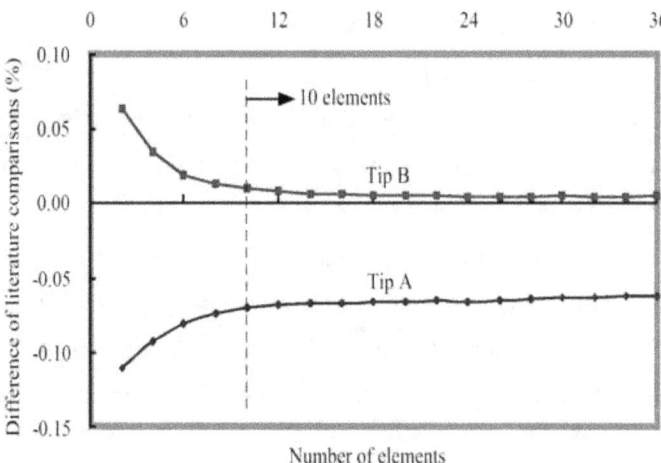

Figure 4.2: Determination of the number of element for crack surface

4.1.2 Determination of the number of element for outer boundary

It was pointed out in the previous example that the accurate calculate SIFs with 10 discontinuous quadratic elements for crack surface, and this section, the number of elements for outer boundary was determinate by the 2D finite domain problem of bi-material, such as follows.

(i)   A rectangular plate

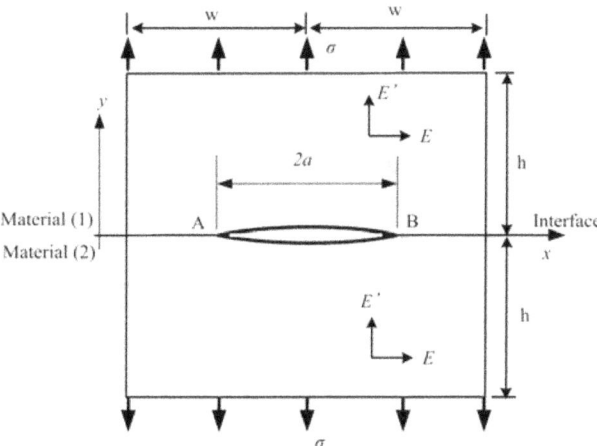

Figure 4.3: An interfacial crack within finite rectangular plate of bi-material.

We analyze a central interfacial crack problem as shown in Figure 4.3 in a finite plate of isotropic bi-material. The static tensile loading σ is applied on the upper and the lower boundary of the plate. The crack has a length, $2a$, h = w, a/w = 0.4, the Poisson ration for both material (1) and (2) are the same, i.e. $v_{(1)} = v_{(2)}$ = 0.3, and the ratio of the Young's module $E_{(1)}/E_{(2)}$ is assumed to be equal to 1. A plane stress condition is assumed. The interfacial crack surface ($2a$) was meshed with 20 discontinuous quadratic elements, and the number of quadratic elements were used to mesh the outer boundary is to vary.

Table 4.2 Calculation for SIFs with several values of the number of elements for rectangular plate. (a/w = 0.4)

Number of elements	$K_I/\sigma\sqrt{\pi a}$		
	Present	Isida (1976)	Diff. (%)
4	1.239146		-1.90344
8	1.208771		0.594468
12	1.215494		0.041624
16	1.215862		0.011369
20	1.216634	1.216	-0.05214
24	1.216816		-0.0671
28	1.216930		-0.07649
32	1.216983		-0.0808
36	1.217013		-0.08328
40	1.217030		-0.08473

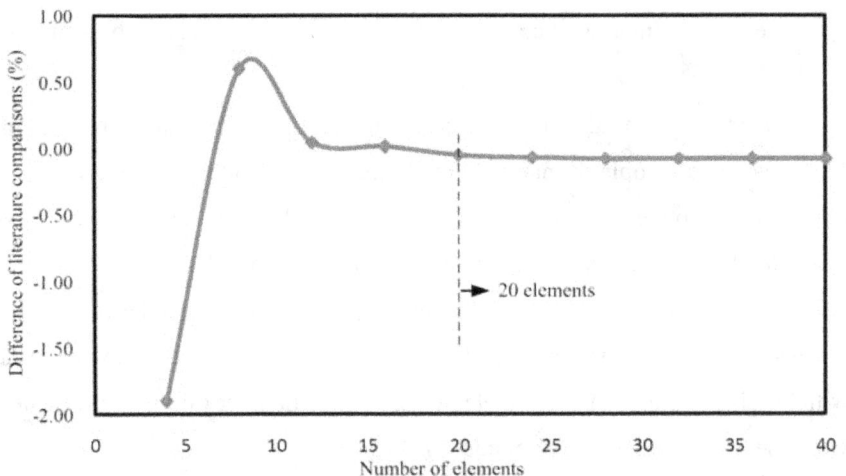

Figure 4.4: Determination of the number of element for out boundary
(rectangular plate)

The stress intensity factors are listed in Table 4.2 for several values of the number of elements, and are compared with the analytical solutions obtained by

Isida, 1976. As shown in Figure 4.4, the curve becomes gentle after 20 quadratic elements; hence we can find that the results between analytical and numerical approaches are very close when outer boundary of the rectangular plate with the number of quadratic elements is equal to 20. It is obvious that 20 or more continuous quadratic elements are suitable for meshing of outer boundary (for rectangular plate).

(ii) A Brazilian disc

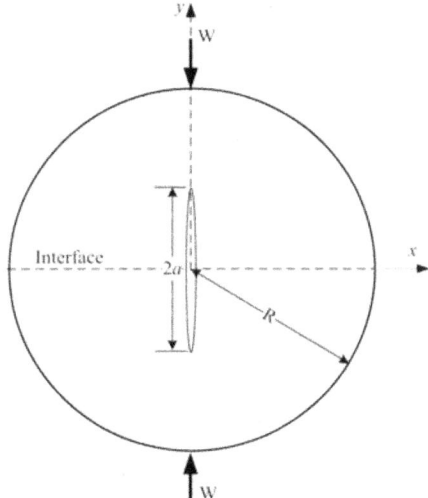

Figure 4.5: A vertical crack within Brazilian disc of bi-material.

It is a vertical crack problem as shown in Figure 4.5 in a Brazilian disc of isotropic bi-material. The diametral compressive loading W is applied on the upper and the lower boundary of the circular disc. The geometry of the problem is that of a thin circular disc of radius R with a central crack has a length, $2a$, a/R = 0.5, the Poisson ration for both material (1) and (2) are the same, i.e. $v_{(1)} = v_{(2)}$ = 0.3, and the ratio of the Young's module $E_{(1)}/E_{(2)}$ is assumed to be equal to 1. Again, the crack surface ($2a$) was meshed with 20 discontinuous quadratic elements, while varies the numbers of elements to mesh the outer boundary. A plane stress condition is assumed.

Table 4.3 Calculation for SIFs with several values of the number of elements for Brazilian disc. ($a/R = 0.5$)

Number of elements	$K_I/\sigma\sqrt{\pi a}$		
	Present	Atkinson et al. (1982)	Diff. (%)
12	1.338579		3.491058
16	1.346071		2.950911
20	1.379511		0.539912
24	1.383921	1.387	0.222012
28	1.387560		-0.04036
32	1.386803		0.014175
36	1.386954		0.003317
40	1.387021		-0.00151

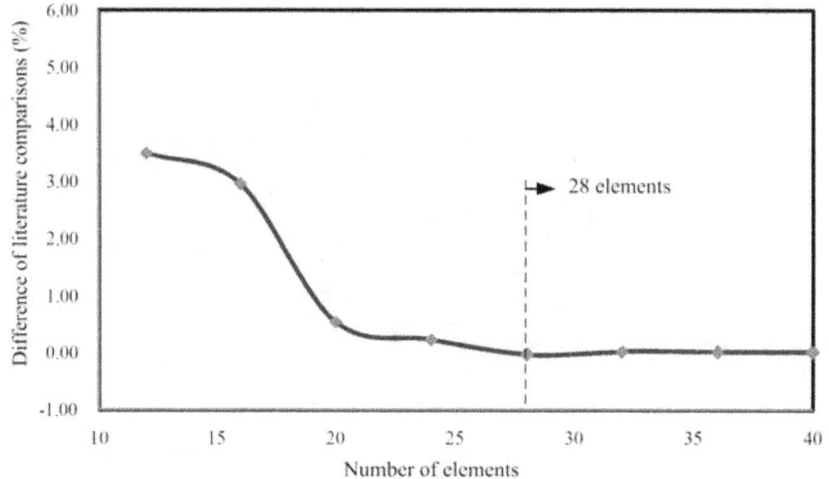

Figure 4.6: Determination of the number of element for out boundary (Brazilian disc)

The SIFs are listed in Table 4.3 for several values of the number of elements, and are compared with the analytical solutions obtained by Atkinson et al., 1982. As shown in Figure 4.6, the curve becomes gentle after 28 quadratic elements; hence we can find that the results between analytical and numerical approaches are close when outer boundary of the circular disc with 28 quadratic elements. It is clear that 28 or more continuous quadratic elements are suitable for meshing

60

of outer boundary (Brazilian disc).

4.2 Determination of stress intensity factors

The number of elements affects accuracy of SIFs that has been understood by example of section 4.1. However, the simulation of crack propagation path is based on obtaining the correct SIFs. In this section, the following numerical examples are presented to verify the SIFs by BEM formulation and to shows the efficiency and versatility of the BEM method for problems related to fracture in isotropic (or anisotropic) bi-material. Unless otherwise noted, in all examples below the crack half-lengths of straight crack is a. Furthermore, all examples are for the case of plane stress in a finite (or infinite) domain of bi-material with the rule of meshing the number of elements according to section 4.1.

4.2.1 Horizontal crack in material (1)

A horizontal crack under a uniform pressure P is shown in Figure 4.7. The crack has a length, $2a$, and is located at a distance, d, to the interface. The Poisson ration for both material (1) and (2) are the same, i.e. $v_{(1)} = v_{(2)} = 0.3$, while the ratio of the shear module ratio $G_{(2)}/G_{(1)}$ varies. A plane stress condition is assumed. In order to calculate the SIFs at the crack tip A or B, 20 quadratic elements were used to discretize the crack surface. The results are given in Table 4.4 for various values of the shear module ratio. They are compared to the results given by Isida and Noguchi, 1983 with using a body force integral equation method and those by Yuuki and Cho, 1989 with using a multi-domain BEM formulation. As can be observed from this table, the results compare quite well.

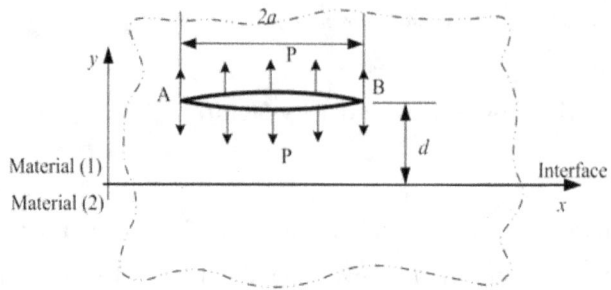

Figure 4.7: A horizontal crack under uniform pressure within material (1) of an infinite bi-material.

Table 4.4 Comparison of the SIFs (horizontal crack)

$\frac{G_{(2)}}{G_{(1)}}$	$\frac{d}{2a}$	Present	Isida et al. (1983)	Diff. (%)	Yuuki et al. (1989)	Diff.(%)
			$K_I/(p\sqrt{\pi a})$ of tip A (or B)			
0.25	0.05	1.476	1.468	-0.57	1.468	-0.54
0.25	0.5	1.198	1.197	-0.09	1.197	-0.12
2.0	0.05	0.871	0.872	0.14	0.869	-0.17
2.0	0.5	0.936	0.935	-0.06	0.934	-0.16
			$K_{II}/(p\sqrt{\pi a})$ of tip A (or B)			
0.25	0.05	0.285	0.286	0.35	0.292	2.50
0.25	0.5	0.071	0.071	0.70	0.072	1.67
2.0	0.05	-0.088	-0.087	-1.38	-0.085	-4.01
2.0	0.5	-0.023	-0.024	2.50	-0.023	-3.54

4.2.2 Vertical crack in material (1)

Consider a vertical crack in material (1) and subjected to far-field horizontal stresses as shown in Figure 4.8. The crack has a length, $2a$, and is located at a distance, d, to the interface. The Poisson ration for both material (1) and (2) are the same, i.e. $v_{(1)}$ =0.35, $v_{(2)}$ = 0.3, and the ratio of the shear module are the same, i.e. $G_{(2)}/G_{(1)}$ = 23.077, while the ratio of the crack length and located at a distance (d/a) varies. A plane stress condition is assumed.

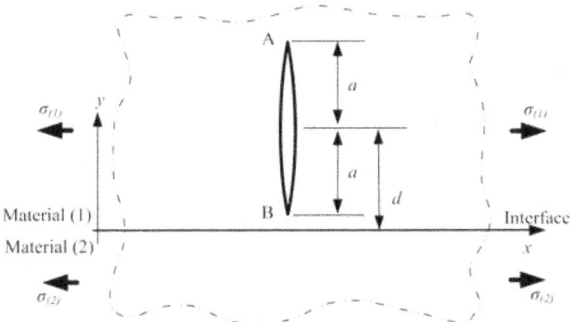

Figure 4.8: A vertical crack in material (1) an interface under far-field stresses.

In order to calculate the SIFs at the crack tip A or B, 20 quadratic elements were used to discretize the crack surface. The results are given in Table 4.5 for various values of the crack length and located at a distance. They are compared to the results given by Isida and Noguchi (1983) using a body force integral equation method and those by Cook and Erdogan (1972) using a Wiener-Hopf technique and an asymptotic analysis. As can be observed from this table, the results compare quite well.

Table 4.5 Comparison of the SIFs (vertical crack)

$\dfrac{d}{a}$	Present	Isida et al. (1983)	Diff. (%)	Cook et al. (1972)	Diff. (%)
		$K_I/(\sigma\sqrt{\pi a})$	of tip A		
1.00	0.882	0.883	0.11	0.883	0.14
2.00	0.961	0.962	0.10	0.962	0.10
5.00	0.995	0.993	-0.17	0.993	-0.17
10.00	0.996	0.998	0.24	0.998	0.24
		$K_I/(\sigma\sqrt{\pi a})$	of tip B		
1.00	-----------	-----------	--------	-----------	---------
2.00	0.935	0.935	-0.04	0.935	-0.04
5.00	0.996	0.991	-0.43	0.991	-0.44
10.00	1.003	0.998	-0.51	0.998	-0.51

4.2.3 Vertical crack intersecting an interface

Consider a vertical crack intersecting an interface and subjected to far-field

63

horizontal stresses as shown in Figure 4.1. The horizontal far field stresses applied in materials (1) and (2) are, respectively, $\sigma_{(1)}$ and $\sigma_{(2)}(=\sigma_{(1)}G_{(2)}/G_{(1)})$. The Poisson ratio $\nu_{(1)}$ and $\nu_{(2)}$ are equal to 0.3, and the shear modulus ratio $G_{(2)}/G_{(1)}$ is assumed to vary. The distance of the crack tip A and B to the interface are the same, i.e. $d_{(1)}=d_{(2)}=a$, the half-length of the crack. Again, a plane stress condition is assumed and 20 quadratic elements were used to discretize the crack surface. The SIFs at the crack tips A and B are listed in Table 4.6 for several values of the shear modulus ratio, and are compared to those given by Isida and Noguchi, 1983. Again, the results between the two numerical analyses compare quite well.

Table 4.6 Comparison of the SIFs (vertical crack intersecting an interface)

$\dfrac{G_{(2)}}{G_{(1)}}$	$K_I/(\sigma\sqrt{\pi a})$ of tip A			$K_I/(\sigma\sqrt{\pi a})$ of tip B		
	Isida and Noguchi (1983)	Present	Diff. (%)	Isida and Noguchi (1983)	Present	Diff. (%)
0.1	1.062	1.0629	-0.08	1.153	1.1539	-0.08
0.3	1.015	1.0157	-0.07	1.064	1.0639	0.01
0.5	1.000	1.0007	-0.07	1.028	1.0273	0.07
0.8	0.997	0.9975	-0.05	1.006	1.0047	0.13

4.2.4 Interfacial horizontal crack in an infinite anisotropic bi-material

An interfacial crack along the x-axis of length $2a$ is shown in Figure 4.9. The crack surface is under a uniform pressure P and the materials can be either isotropic or anisotropic. Twenty quadratic elements were used to discretize the crack surface and the characteristic length is assumed as $2a$.

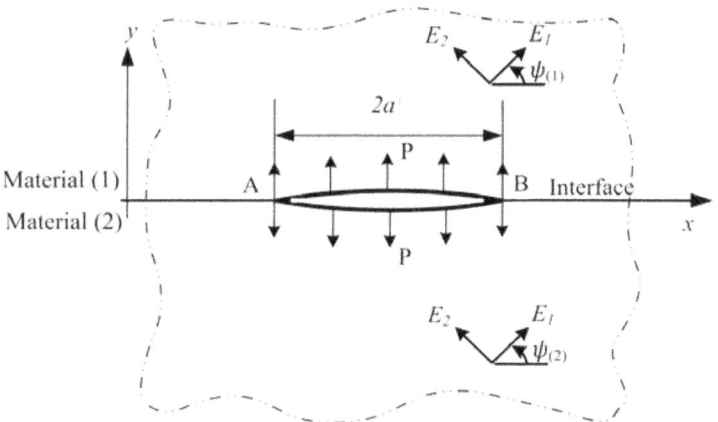

Figure 4.9: Interfacial crack within infinite bi-materials.

The SIFs at the crack tip of an interfacial crack were also calculated for the anisotropic bi-material case. The anisotropic elastic properties in material (1) were assumed to be those of glass/epoxy with $E_1 = 48.26$GPa, $E_2 = 17.24$GPa, $G_{12} = 6.89$GPa, $v_{12} = 0.29$. For material (2), a graphite/epoxy with $E_1 = 114.8$GPa, $E_2 = 11.7$GPa, $G_{12} = 9.66$GPa, $v_{12} = 0.21$ was selected (Dwyer and Pen, 1995). The material axis E_1 in material (1) and material (2) makes angles $\psi_{(1)}$ and $\psi_{(2)}$ respectively, with respect to the horizontal direction (Figure 4.9.). While the material axis E_1 in material (2) was assumed to be along the horizontal direction (i.e. $\psi_{(2)} = 0°$), the E_1-axis in material (1) makes an angle $_{(1)}$ with respect to the horizontal direction. The interfacial SIFs at crack tip B obtained by the present method are listed in Table 4.7 and compared to the exact solutions proposed by Wu, 1990. A very good agreement is found between the numerical analysis and the exact solution.

Table 4.7 Comparison of the SIFs for infinite anisotropic problem
$$\psi_{(2)} = 0°$$
(interfacial crack)

$\psi_{(1)}$	$K_I/(p\sqrt{\pi a})$ of tip A (or B)			$K_{II}/(p\sqrt{\pi a})$ of tip A (or B)		
	Wu, (1990)	Present	Diff. (%)	Wu, (1990)	Present	Diff. (%)
0	1.0000	1.0053	-0.53	-0.0382	-0.0381	0.26
30	0.9968	1.0006	-0.38	-0.0349	-0.0350	-0.29
45	0.9965	1.0001	-0.36	-0.0318	-0.0319	-0.31
60	0.9971	1.0010	-0.39	-0.0290	-0.0292	-0.69
90	1.0000	1.0054	-0.54	-0.0264	-0.0265	-0.38

4.2.5 Interfacial horizontal crack in finite bi-materials

Table 4.8 Elastic properties for the material (1) and (2)

Materials	E (MPa)	E' (MPa)	v'	G'
Material (1)	100	50	0.3	10.009
Material (2)				
(i)	100	45	0.3	9.525
(ii)	100	40	0.3	9.010
(iii)	100	30	0.3	7.860
(iv)	100	10	0.3	4.630

The example is included as a comparison with the literature in order to demonstrate the accuracy of BEM approach for an interfacial crack in anisotropic bi-material plate. The geometry is that of rectangular plate and is shown in Figure 4.3. For the comparison, crack length is taken as $2a=2$, h = 2w, a/w = 0.4, and static tensile loading σ is applied on the upper and the lower boundary of the plate. Plane stress condition is assumed. The anisotropic elastic properties for the material (1) and (2) are given in Table 4.8. The normalized complex stress intensity factors at crack tip A or B is listed in Table 4.9 together with those from Cho et al. (1992), who used a multi-domain BEM formulation and the results from Wünsche et al. (2009) for a finite body. The outer boundary and interfacial crack surface were discretized with 20 continuous and 20 discontinuous quadratic elements, respectively. It is obvious from Table 4.9 that these are very close to those obtained by the other researchers.

Table 4.9 Comparison of the Normalized complex SIFs for finite anisotropic problem   (interfacial crack)

Material (2) E'/E	$|K|/(\sigma\sqrt{\pi a})$ of tip A (or B)				
	Cho et al. (1992)	Wünsche et al. (2009)	Diff. (%)	Present	Diff. (%)
(i): 0.45	1.317	1.312	0.38	1.3132	0.29
(ii): 0.40	1.337	1.333	0.30	1.3351	0.14
(iii): 0.30	1.392	1.386	0.43	1.3922	-0.01
(iv): 0.10	1.697	1.689	0.47	1.6968	-0.06

*$|K| = \sqrt{K_I^2 + K_{II}^2}$

However, the proposed BEM program is not only very accurate but also very fast. The calculation time for each example is typically 30s by BEM code with a Personal Computer, PC (Intel Pentium 4, 2.4 GHz CPU, 1GB RAM).

4.3 Verification of the crack initiation angle

As mentioned previously, the proof of the SDBEM program can accurately calculate the stress intensity factor and get satisfactory numerical results. In this section, we may now proceed to verify the validity of the SDBEM program by calculating the crack initiation angle, and compared the following three experimental results for illustration.

4.3.1 Centrally straight-notched Brazilian disc (CSNBD)
-by Naser A. Al-Shayea (2005)

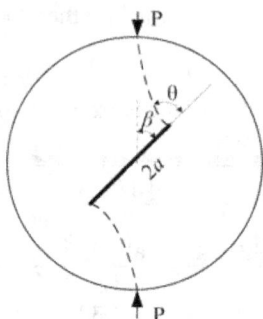

Figure 4.10: CSNBD specimen of limestone rock by Naser A. Al-Shayea (2005)

Naser A. Al-Shayea (2005), present the centrally straight-notched Brazilian disk (CSNBD) specimens of limestone rock, with various crack inclinations, were used to study the crack propagation behavior under mixed mode I–II loading. These specimens have a central crack and were loaded with diametrical compression (as shown in Figure 4.10). The limestone rock discs 98mm and 84mm in diameter and 22mm in thickness with 30mm notch. Experimentations were made at ambient conditions, at high confining pressure (σ_c) of 28 MPa, and at high temperature of 116 °C. Other details of the experimental setup were reported Al-Shayea et al. (2000). Among others, the vertical load and the crack deformation were continuously recorded during the test using a computerized data logger. Crack deformation was monitored only at ambient conditions. The crack inclination angle (β) between the crack plane and the diametrical compression was varied. Results are compared with the theoretical analysis of crack propagation under mixed mode I–II loading condition based on the maximum tangential stress (σ-criterion). Figure 4.11 shows the variation of the crack initiation angle θ with the crack inclination angle β determined numerically and experimentally. A good agreement is found between the experimental results of Naser A. Al-Shayea (2005) and our numerical predictions.

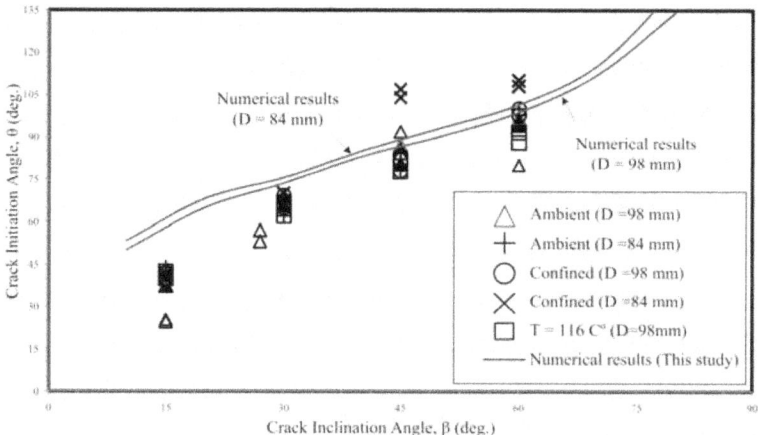

Figure 4.11: Variation of crack initiation angle θ with the crack angle β. CSNBD specimen of limestone rock subjected to diametral compression.

4.3.2 Prismatic specimens of kaolonite clay –by Vallejo (1987)

Vallejo (1987) conducted uniaxial compression tests on cracked prismatic specimens of kaolonite clay76.2×76.2×25.4 mm in size containing a central crack 24.9 mm in length (as shown in Figure 4.12).

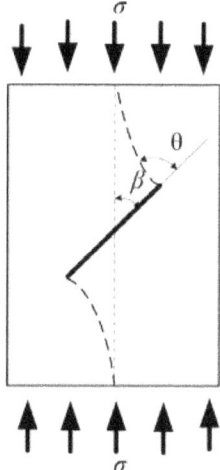

Figure 4.12: Prismatic specimens of kaolonite clay –by Vallejo (1987)

Several tests were carried out by varying the crack inclination angle (β) between the crack plane and the compressive stress. Figure 4.13 shows a comparison between the crack initiation angles (θ) measured experimentally and those predicted numerically. A good agreement is found between the experimental results of Vallejo (1987) and our numerical predictions.

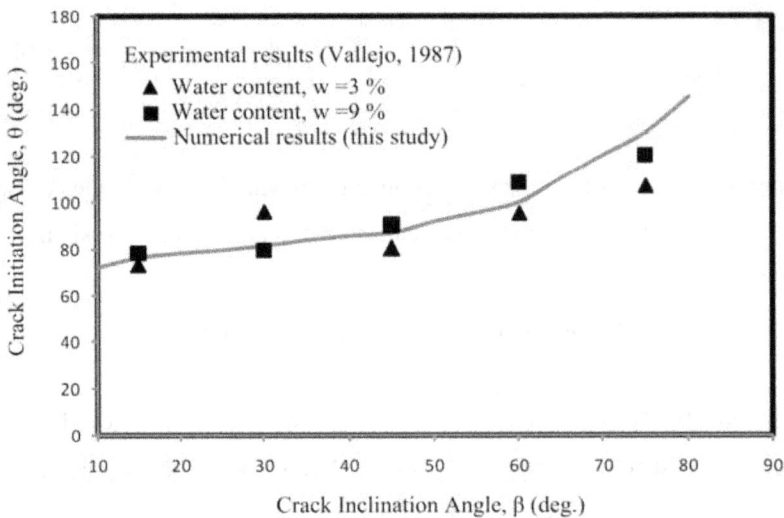

Figure 4.13: Variation of crack initiation angle θ with the crack angle β. Prismatic sample of kaolinite clay subjected to uniaxial compression.

4.3.3 Isotropic Plexiglass sheets -by Erdogan and Sih (1963)

Erdogan and Sih (1963) conducted uniaxial tension tests on isotropic Plexiglass sheets 229×457×4.8 mm in size containing a 50.8 mm long central crack (as shown in Figure 4.14). The crack orientation angle (β) between the crack plane and the tensile stress was varied. Figure 4.15 shows the variation of the crack initiation angle (θ) with the crack inclination angle (β) determined numerically and experimentally. Again, a good agreement is found between the two approaches.

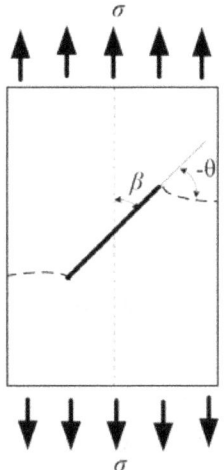

Figure 4.14: Isotropic Plexiglass sheets by Erdogan and Sih (1963)

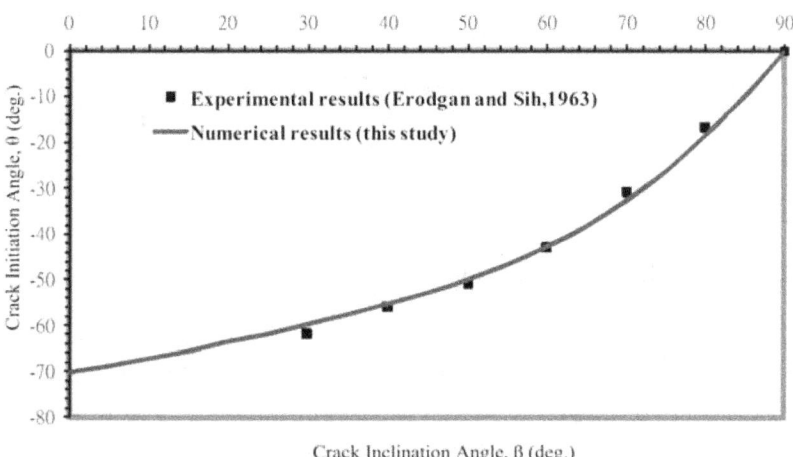

Figure 4.15: Variation of crack initiation angle θ with the crack angle β.
Plexiglass plate subjected to uniaxial tension.

4.4 Modeling crack propagation path

In the previous section, we prove that the proposed SDBEM program can not only accurately calculate stress intensity factors of crack tip but also obtain a reasonable initiation angle. As far as the simulation for the crack propagation path is concerned, we pay more attention to the accuracy of the numerical results after crack extension.

4.4.1 Stress intensity factor of extension crack tip

(i) An interface kinked crack (extension crack) in infinite bi-material

We consider an interfacial crack in infinite bi-materials and subjected to far-field tensile stresses. The interfacial crack length is 2c and a new crack length which extends from tip B along the initial direction at an angle of 45 degree within material (1) is 2a, such as shown in Figure 4.16.

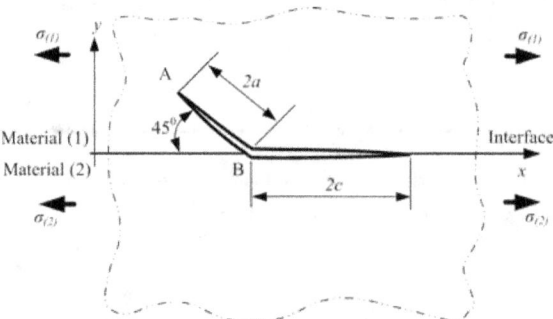

Figure 4.16: Interfacial kinked crack within infinite bi-materials.

The far-field tensile stresses applied in materials (1) and (2) are, respectively, $\sigma_{(1)}$ and $\sigma_{(2)} = (= \sigma_{(1)}G_{(2)}/G_{(1)})$. The new tip is A of extension crack within material (1). The Poisson ratio $v_{(1)}$ and $v_{(2)}$ are assumed to be equal to 0.3. Plane stress condition is assumed. The Young's module in materials (1) is 1GPa and according the shear module ratio with material (1) and material (2) ($G_{(1)}/G_{(2)} = 0.25$) we can get the Young's module of material (2). Then according equation ($G = (E/2(1+v))$), we can get the relation between extension length and stress intensity factors. After numerical analysis, we compared with Isida and Noguchi, 1983 using a body integral equation force method, as shown in Table 4.10, the results between the two numerical analyses compare quite well.

Table 4.10 Comparison of the SIFs (interfacial kinked crack)

$\frac{a}{c}$	$K_I/\left(\sigma\sqrt{\pi a}\right)$ of Tip A			$K_{II}/\left(\sigma\sqrt{\pi a}\right)$ of Tip A		
	Isida and Noguchi (1983)	Present	Diff. (%)	Isida and Noguchi (1983)	Present	Diff. (%)
0.2	0.733	0.759	-3.54	0.631	0.624	1.10
0.5	0.708	0.730	-3.10	0.623	0.612	1.76
1.0	0.683	0.701	-2.63	0.610	0.603	1.14
1.5	0.669	0.685	-2.39	0.601	0.594	1.16

(ii)   An edge crack in a semi-infinite domain

We consider an edge crack in a semi-infinite isotropic domain and subjected to horizontal far- field tensile stresses. The propagation path of edge crack as shown in Figure 4.17. Here, each extension length is a, the entire extension length is c, and number of extension times is N, plane stress condition is assumed.

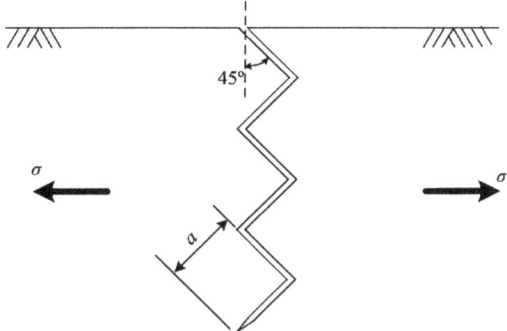

Figure 4.17: An edge crack in a semi-infinite domain

The horizontal far field stresses applied in materials is σ. The Poisson ratio v is 0.3. The entire extension length and each extension length are the same, i.e. c = Na. Again, a plane stress condition is assumed and 20 quadratic elements were used to mesh the crack surface. The SIFs at the each extension of new crack tips as listed in Table 4.11, and can be compared to those given by Isida, 1979. Again, the results between the numerical analyses and the exact solution compare quite well. By the way, the proposed SDBEM program which assumed elastic properties to come near zero with on material (1) (or (2)) can also approach to the problem of semi-infinite domain.

Table 4.11 Comparison of the SIFs (interfacial kinked crack)

N	$K_I/(\sigma\sqrt{\pi c})$			$K_{II}/(\sigma\sqrt{\pi c})$		
	Isida (1979)	Present	Diff. (%)	Isida (1979)	Present	Diff. (%)
1	0.705	0.706	-0.13	0.365	0.365	0.13
2	0.703	0.706	-0.46	0.364	0.366	-0.68
3	0.704	0.708	-0.59	0.360	0.362	-0.42
4	0.704	0.712	-1.07	0.355	0.356	-0.24
5	0.708	0.715	-1.00	0.349	0.351	-0.65
6	0.707	0.718	-1.58	0.348	0.348	0.07

Chapter 5

Experimental Investigation

After accuracy checking, we take cement and gypsum to design the bi-material experiment which provides first a detailed overview of the test program in this chapter. Then, it gives a comprehensive description of testing equipment used to make and test for the bi-material specimens. Further, the experimental procedure is presented, the results are exposed and the accuracy at each step is finally discussed.

5.1 Objective and program

The main purpose of this testing program is that to study the fracture initiation θ0 and propagation within bi-material specimens and to measure the maximum load W_f at failure for bi-material specimen. The flowchart of this experimental investigation can be shown in Figure 6.1. There are two kinds of experimental investigation: the experiment of materials properties to determine the elastic properties of the materials, and the fracture experiment to measure the fracture propagation within bi-material specimens. The determination of elastic properties of the materials which to include the Brazilian test method (indirect tensile strength determination) and the test of uniaxial compression (uniaxial compressive strength determination) have been suggested by ISRM (International Society for Rock Mechanics) used to make the specimens. And the fracture experiment, we design the cement/gypsum bi-material specimens of Brazilian disc (Cracked Straight Through Brazilian Disc, CSTBD) to measure the fracture initiation θ0 and propagation of bi-material specimens. On the other hand, the numerical results are calculated by BEM formulation which input boundary condition of specimen geometry and elastic properties of materials should obviously be the same as that in the experiments. The numerical results and the experimental results are finally discussed.

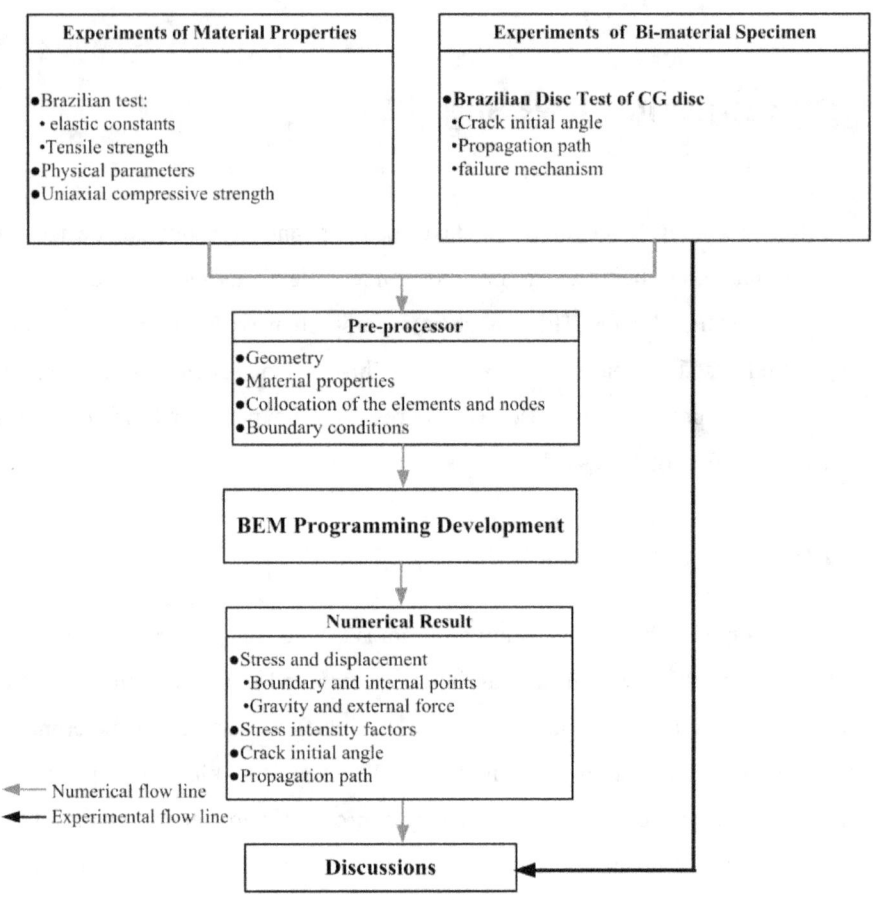

Figure 5.1: Flowchart of this experimental investigation.

5.2 Specimens production

In this experimental investigation, the materials of specimens are using Type I Portland cement, also known as general cement, and gypsum. To determine the elastic properties of the cement and gypsum, two samples of each kind have been instrumented by strain gage rosettes and tested accordingly to the ISRM

76

suggested method. The specimens of material properties were grouped as shown in Table 5.1.

Table 5.1 The specimens grouping of elastic constants

	cement	gypsum
Brazilian Disc	CD-01~03	GD-01~03
Cylindrical specimens	CC-01~03	GC-01~03

In this experiment, two different steel mold templates are used to prepare the Brazilian disc specimens and bi-material Brazilian specimens of cement and gypsum (Figure 5.2). The Brazilian disc specimens are used to determine the elastic constants of cement and gypsum. The Brazilian test is conducted by gluing a 45° strain gage rosette on the center of each disc.

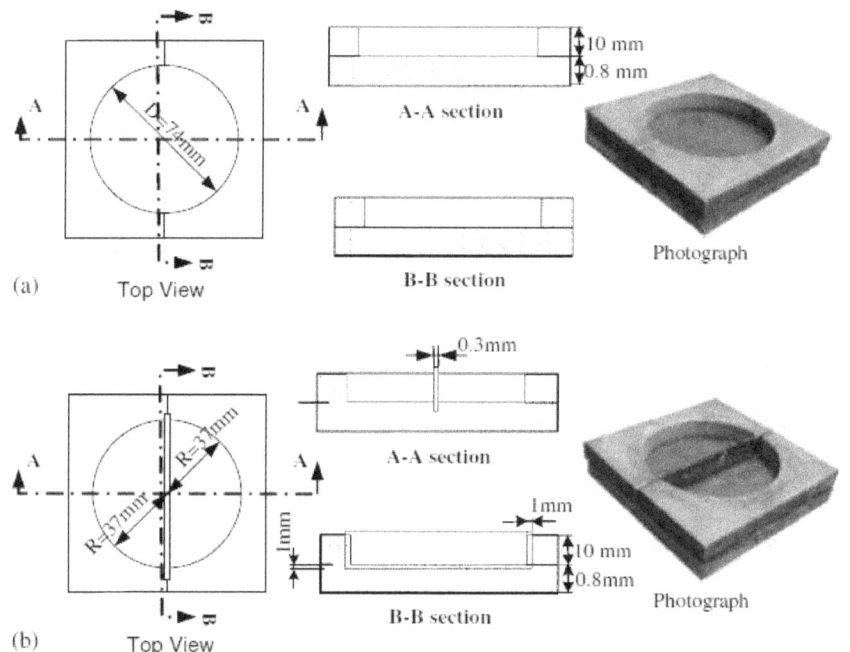

Figure 5.2: Two steel mold templates for: (a) Brazilian disc specimen and (b) bi-material Brazilian specimen.

Afterward, the CSTBD of bi-materials have been assembled utilizing the two previous rocks. The category consisted of two halves of each rock put together to form cement/gypsum specimens (CGD1 toCGD7). The study put a special focus on the influence of the crack orientation β and interface orientation λ on the crack propagation path. For each specimen, the maximum load was recorded as well.

Table 5.2 CSTBD bi-material specimens, type CGD.

	No.	$\beta=0°$	$\beta=45°$	
$\lambda=90°$	1	CGD1-1	CGD2-1	
	2	CGD1-2	CGD2-2	
	3	CGD1-3	CGD2-3	
$\lambda=135°$	1	CGD3-1	CGD4-1	
	2	CGD3-2	CGD4-2	
	3	CGD3-3	CGD4-3	
$\lambda=45°$	1	CGD3 series of the same symmetry	CGD5-1	
	2		CGD5-2	
	3		CGD5-3	
$\lambda=0°$	1	CGD6-1	CGD7-1	
	2	CGD6-2	CGD7-2	
	3	CGD6-3	CGD7-3	

In this study, the effects of diameter and thickness of the bi-material CSTBD specimens were isolated. In fact, all the bi-material CSTBD specimens were cut in such a way to exhibit a diameter equal to 74 mm and a thickness of 12 mm. Once all the tests were performed, both sides of samples were photographed, and the data were plotted. These data are also used as inputs in a BEM program, to determine the SIFs, fracture initiation θ_0 and propagation path for the bi-material specimens.

5.2.1 Strain gage installation

As mentioned before, two samples of the cement and gypsum were cut accordingly to the ISRM recommendation. In this section, it has been decided to use these samples to measure the elastic properties of both rocks. For this reason, it was necessary to equip the discs with strain gages. The installation tools of strain gage as shown in Figure 5.3.

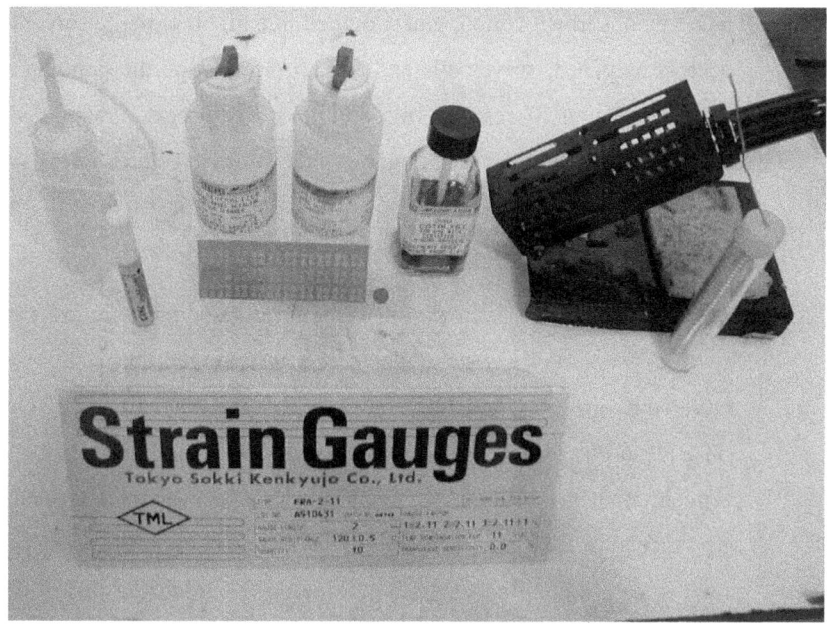

Figure 5.3: The strain gage installation tools

The strain gages (TML, type FRA-2-11, Tokyo Sokki Kenkyujo Co., Ltd. by Japan.) used to test the elastic properties of the cement and gypsum rocks are 45° rosettes (as shown in Table 5.3). Each rosette is composed of three 120-ohms strain gages, measuring the strain horizontally, vertically and in between. The gage factor is 2.11 at 23 °C. The TML Strain gauge adhesive procedure is as follows.

Table 5.3 The TML Strain gages type (45° 3-element Rosette)

Gauge pattern	Type	Gauge length (mm)	Gauge width (mm)	Backing length (mm)	Backing width (mm)	Resistance (Ω)
FRA-2-11	FRA-2-11	2	0.9	Ø7		120

(i)   Surface treatment (see Figure 5.4(a))

Like drawing a circle with sandpaper (#300 or so), polish the strain gage bonding site in a considerably wider area than the strain gage size. And using an absorbent cotton, gauze dipped in a highly volatile solvent such as acetone which dissolves oils and fats, strongly wipe the bonding site in a single direction to remove oils and fats. Reciprocated wiping does not clean the surface. After cleaning, mark the strain gage bonding position.

(ii)   Adhesive curing (see Figure 5.4(b))

Once the surface is clean and smooth, alignment marks are put on the sample using a 4H pencil. The gage is removed from its transparent envelop by grasping its edge with tweezers, and placed bonding side down on a clean glass plate. A cellophane tape is then used as a carrier to aid in repositioning the strain gage. To place the tape over the gage, one end is first tacked to the glass plate behind the gage, and then wiped forward. The tape is carefully lifted at a shallow angle, bringing the gage up with it. Holding the tape at a shallow angle, the assembly is wiped onto the specimen surface. Then, one end of the tape is lifted up at a shallow angle until the gage is free of specimen surface. The loose end of the tape is then set on the specimen so the gage lies flat with the bonding side exposed. Once the specimen and back of the gage are coated with the prepared adhesive, the tucked-over end of tape is lifted and bridged over the adhesive at approximately a 30 degree angle. A silicon gum pad and backup plate are placed over the gage installation. This allows the clamping force to be exerted evenly over the gage. The clamping kept in position during curing time. After 1 minute, the clamp is removed and the tape is pulled back, peeling off the surface. The gage is ready for wiring.

(iii) Wiring operation (see Figure 5.4(c))

Six wires were soldered on each terminal piece, each couple corresponding to one strain gage. This operation is most delicate part of the gage installation; any lack of attention might lead to the destruction of the rosette, caused either by the contact between the soldering tip and the gage, or by bridging two adjacent wires with rosin. At the other end, the wires are connected to three slots. Each slot is then hooked up to one channel on the data acquisition system, subsequently reading the strain in one direction (horizontally, or vertically, or in between).

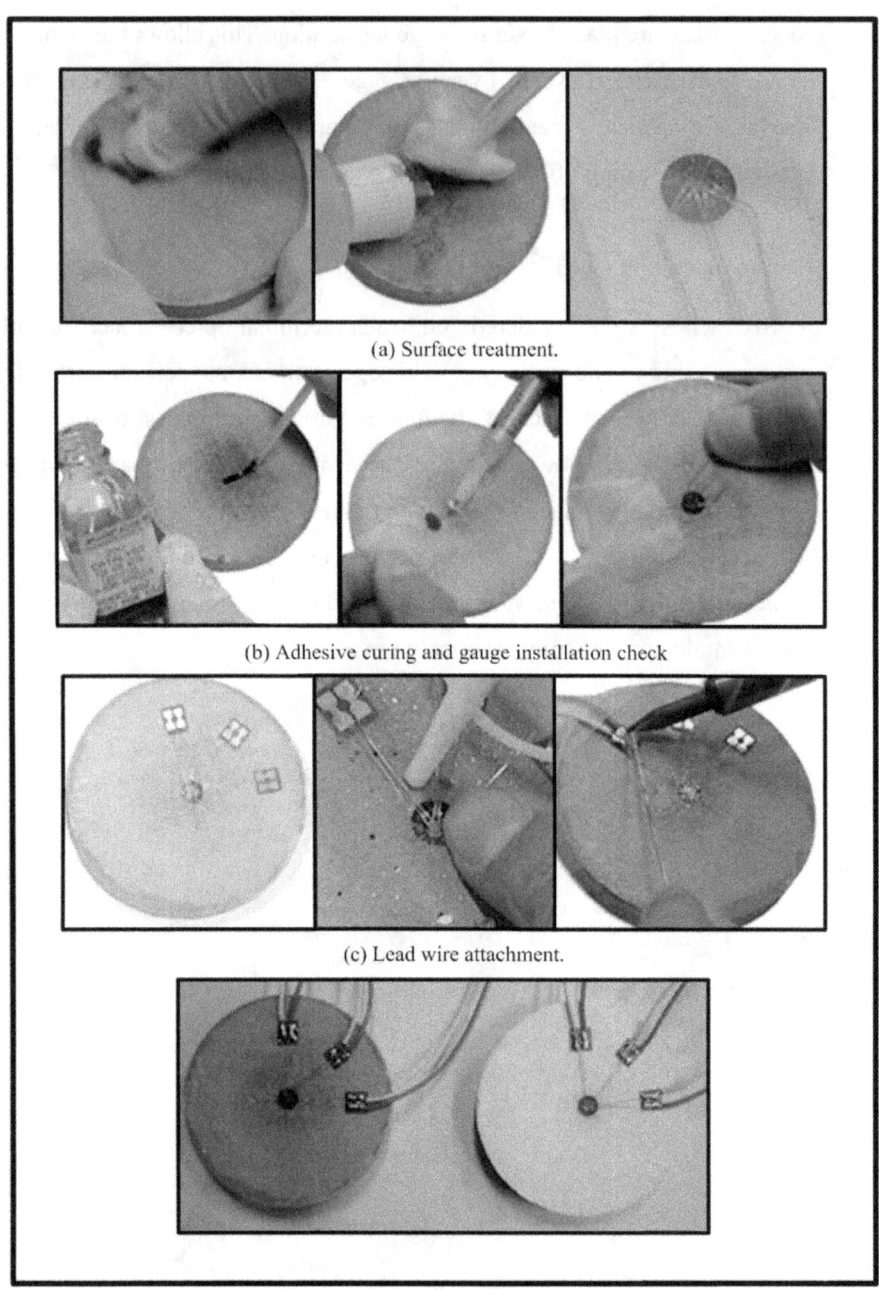

(a) Surface treatment.

(b) Adhesive curing and gauge installation check

(c) Lead wire attachment.

Figure 5.4: Strain gage installation products

5.2.2 Gluing and wrapping samples

To assemble the specimens of bi-material, the two-component glue was used. BOLT A+B is a heavy-duty epoxy paste that offers the advantage of being very efficient on concrete, masonry, and rock (as shown in Figure 5.5). It is also waterproof and chemically resistant. Component A is essentially an epoxy resin, whereas component B is a polyamide hardener.

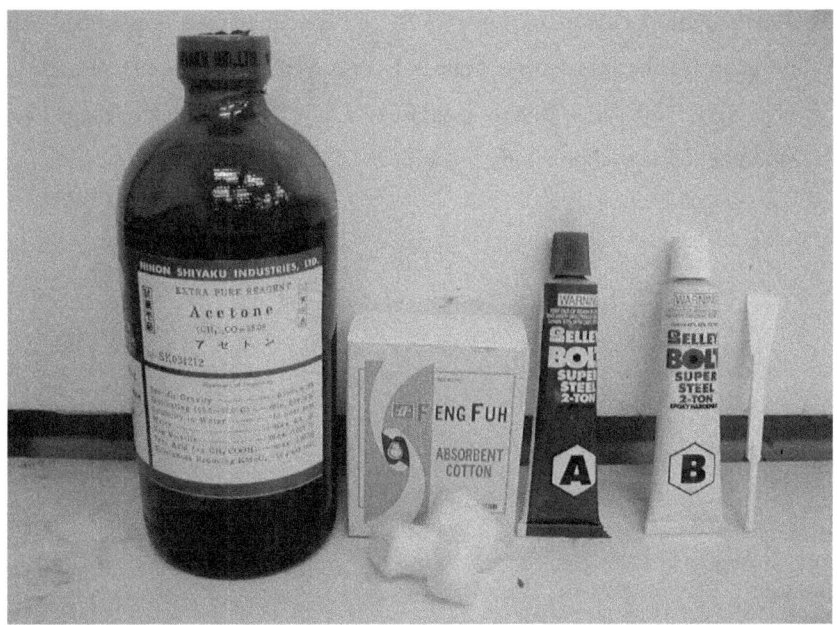

Figure 5.5: Gluing tools of bi-material specimen

To prepare the final mixture, equal amounts of both are taken by the means of knives, and thoroughly mixed on a container (see Figure 5.6(c)). Once the mixture is homogeneous, the gluing operation itself can start. Since AB glue cures within 24 hours, repositioning and adjusting are possible. It is recommended to use the mixture within 2 hours after preparation, for convenience. Then, prior to gluing phase, drafting tape covered all pieces, except in the contact zone where glue is to be applied. This operation is of importance to prevent glue from spreading all over the samples surface. It is obvious that in case glue covers partly the surface of a given specimen, it may guide its behavior with regard to the crack propagation, and consequently induce errors in the results. The contact zone is then treated by the means of degreaser and gauze sponge, in order to eliminate dust, and oil residue deposited (see Figure 5.6(d)(e)). For bi-material specimens, glue is applied over the contact zone, and both halves are gathered and squeezed so that not more than 1 mm of glue stays between the two pieces (see Figure 5.6(f)). The specimen is then laid down flat on fixture, and left at this position for 24 hours coring (see Figure 5.6(i)). After 24 hours curing at 23°C, all specimens are unwrapped. Extra glue popping off the interface is delicately removed by the means of a cutter and if necessary, both sides of the specimens along the contact zone are evened and smoothened with a polishing machine.

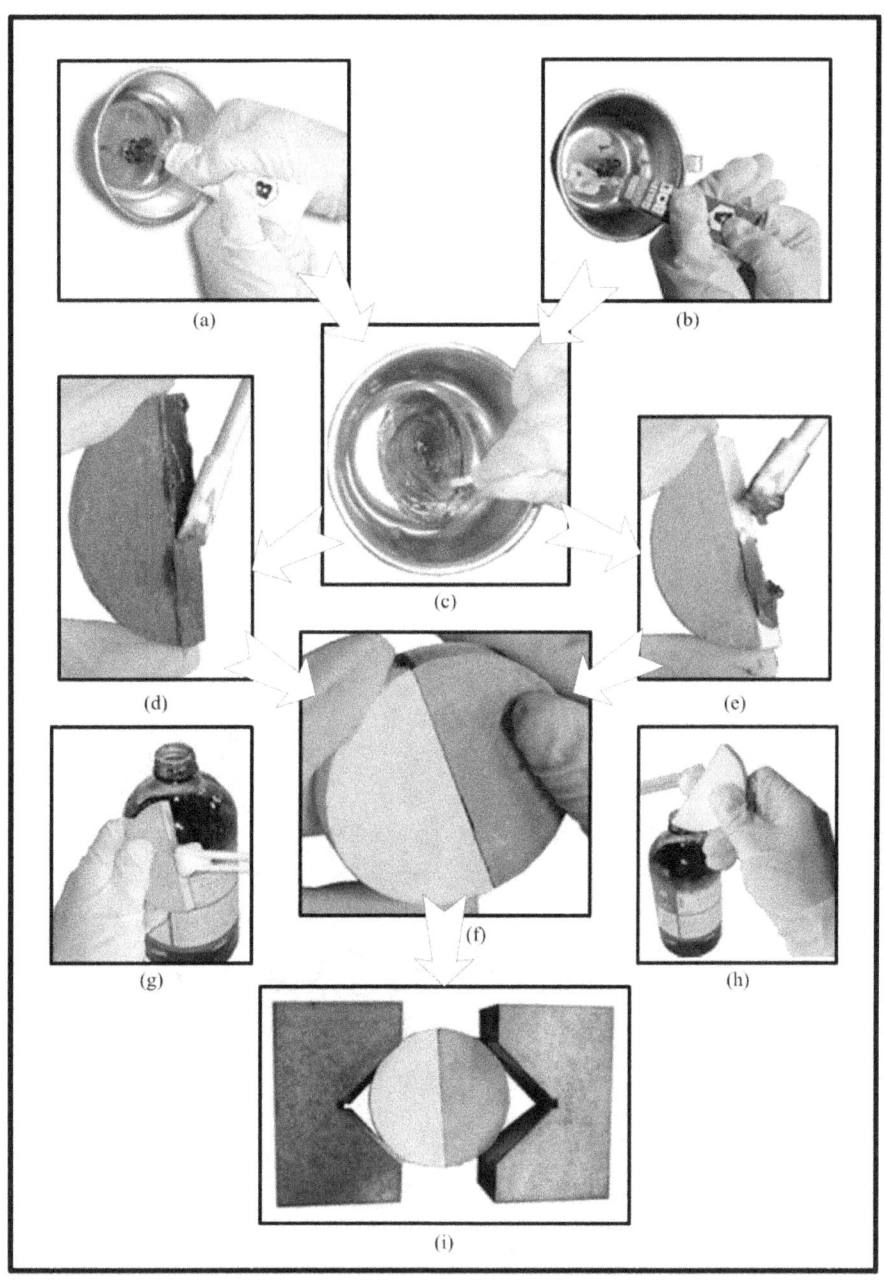

(a)

(b)

(c)

(d)

(e)

(f)

(g)

(h)

(i)

Figure 5.6: Gluing and wrapping samples

5.2.3 Cutting pre-cracks

In this study, the pre-cracks of each specimen are all geometric, length 20 mm, width 2.5 mm of pre-crack. The cutting equipment is utilized to incise a pre-crack in the thin disc, in order to make the CSTBD bi-material specimens. It consists of a Handy Grinder Tool (see Figure 5.7) with variable speed, and is most frequently used as drilling, carving or engraving tool on wood or metal. A circular blade fixed on a Handy Grinder was employed to pierce the pre-crack in each specimen. The diameter of the wheel is approximately 23 mm, and its thickness is equal to 1 mm. This kind of blade is basically designed for metal applications, but it efficiently served in this study to cut through soft rock specimens. Also, to make the pre-crack rigorously conform to the specimens, a jigsaw blade, 1 mm thickness, was chosen to abrade the V-shape tip of the pre-crack.

Figure 5.7: Handy Grinder cutting equipment.

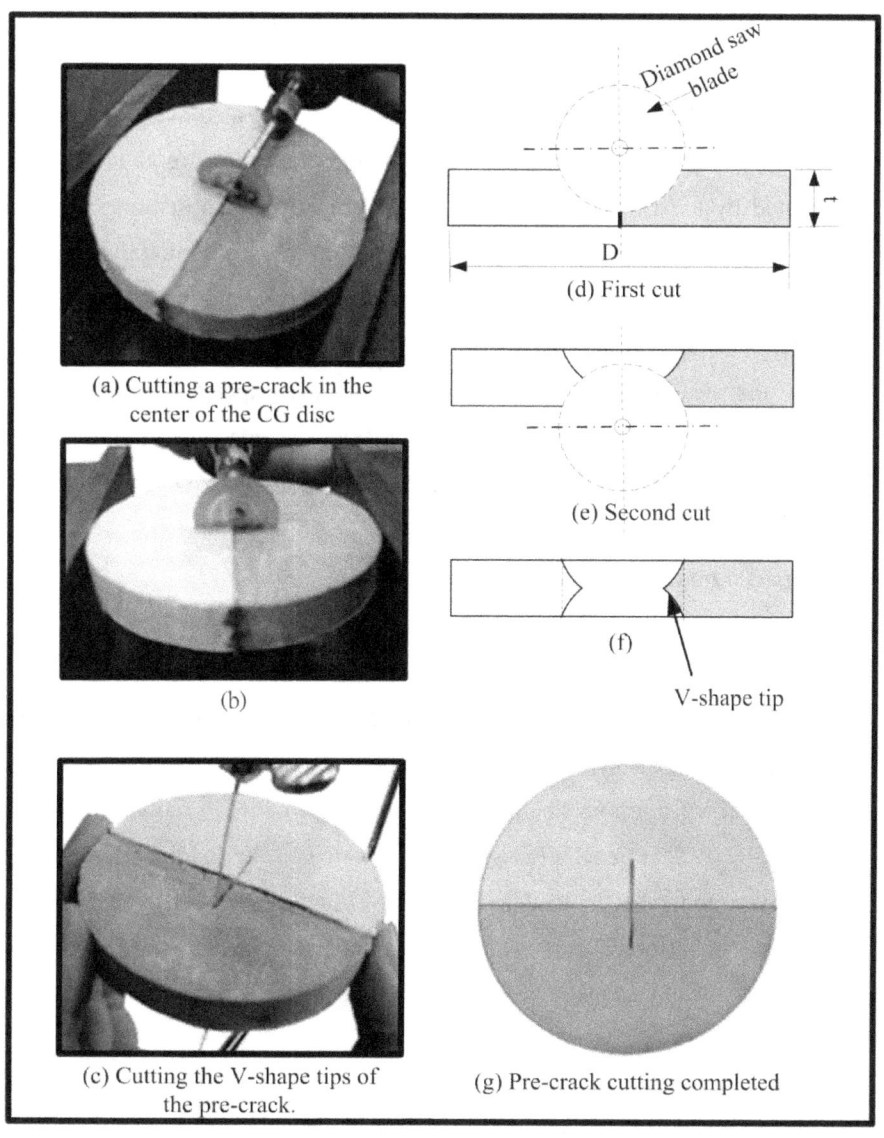

(a) Cutting a pre-crack in the center of the CG disc

(b)

(c) Cutting the V-shape tips of the pre-crack.

Diamond saw blade

(d) First cut

(e) Second cut

(f)

V-shape tip

(g) Pre-crack cutting completed

Figure 5.8: Cutting procedures to make the specimen.

Prior to cutting a pre-crack in the each specimen, the bounds and the orientation of the pre-crack with respect to the horizontal are marked on each specimen. Each specimen is placed between the jaws of a vise (see Figure 5.8(a)), in a horizontal position. The pre-crack direction is recommended to be oriented straightforward. This precaution allows keeping the verticality of the pre-crack under control while carving it into the rock. In order to incise a 20 mm long pre-crack, the blade is only carried 5 to 7 mm into the rock. Then, the sample is turned upside down, and same operation is repeated over. At this phase, the pre-crack is characterized by two V-shape ends that are then gradually eliminated by a 1 mm jigsaw blade (see Figure 5.8(c)). The cutting operation generates excessive dust that covers the marks initially drafted on the specimen surface. Therefore, a small brush is helpful to sweep off the dust, and carry on the operation afterward. This delicate operation is the most time consuming stage of the specimen preparation. It is also the most expensive. Small and thin blades alike the ones used in this study are not designed for cutting through rocks. As a consequence, a new blade has been used for almost each pre-crack.

5.3 Experiments description

Two different types of testing equipment have been used. To determine the elastic properties of the cement and gypsum specimen, and break the bi-material specimens, a loading machine which co-operate with MTS 407 controller was employed, at a 1 mm/min loading rate. Load system configuration as shown in Figure 5.9. A data acquisition system connected to the loading machine was first foreseen to record the load and the strain showed by the strain gage rosettes. This data acquisition system feature a control adapter card (KYOWA DBB-120A, by Japan) inserted into a computer (Computer software is installed InstruNET).

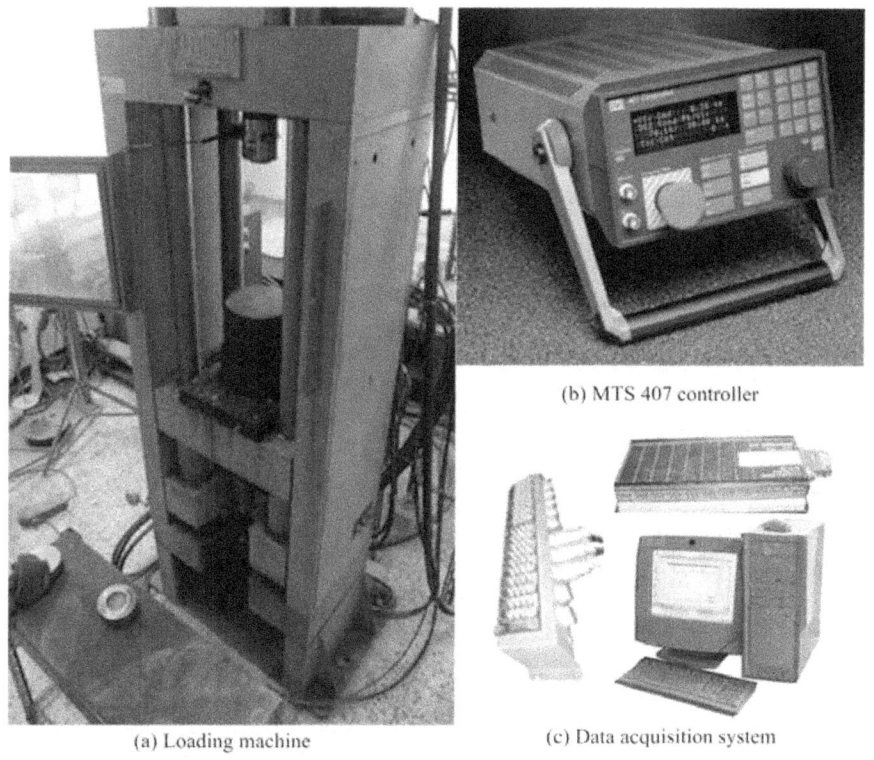

(b) MTS 407 controller

(a) Loading machine

(c) Data acquisition system

Figure 5.9: Loading system and data acquisition system

In order to apply a strip load on the Brazilian samples, and void any kind of premature fracturing, the discs were all placed within steel jaws as described on Figure 5.10. The steel jaws are then positioned on the lower plate of the loading machine and the test is started.

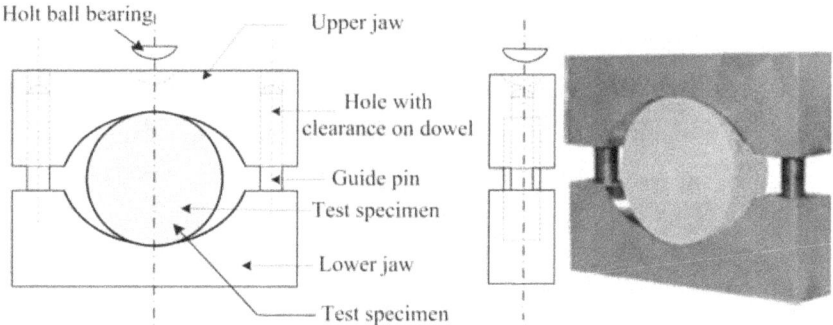

Holt ball bearing

Upper jaw

Hole with clearance on dowel

Guide pin

Test specimen

Lower jaw

Test specimen

Figure 5.10: Steel jaws to apply Brazilian strip load.

5.3.1 Brazilian test

Discs of the two specimens of cement and gypsum were prepared with a thickness-to-diameter ratio of about 0.5. All preparation procedures followed the ISRM suggested methods (Bieniawski and Hawkes, 1978). Two steel jaws were inserted between the discs and the platens of the loading frame in order to apply a strip load instead of a line load. A typewriter film ribbon was placed between the jaws and the specimen to measure the contact angle 2α. Based on measurements, an average arc angle (2α) of 15° was adopted for all calculations in this study. A 45° strain gage rosette was glued at the center of each disc. After preparation, all specimens were loaded up to failure at a constant loading rate of 1 mm/min by using the 100ton loading system.

The procedure used to determine the elastic constants of each one of the two specimens by diametral loading can be summarized as follows:

(i)   The normal strains in the vertical, horizontal, and 45° directions were measured by gluing a 45° strain gage rosette at the center of each disc specimen.

(ii) Two steel-loading jaws were designed to apply a diametral load to each disc specimen over an arc of contact angle (2α).

(iii) Each disc specimen was loaded up to failure.

(iv) The elastic properties were determined from the strain gage measurements by method of Chen, 1996.

The tensile strength of the specimen was also determined by recording the failure load during the Brazilian tests and by using the expression of σ_x (horizontal stress) (Hondros, 1959).

5.3.2 Uniaxial compression test

By definition, the uniaxial compressive strength (UCS) of a specimen is that value of uniaxial compressive stress reached when the specimen fails completely. The compressive strength is usually obtained experimentally by means of a compressive test. As can be imagined, the cylindrical specimen is shortened as well as spread laterally. Method for uniaxial compression test follows the suggested method given by the ISRM (1981). The tests in this study were conducted under diametral conditions. All core samples for this test were drilled perpendicular to bedding, had a minimum length to diameter ratio of 2, and met the strict tolerance limits as specified in the suggested test procedure. In this study, a total of 6 uniaxial compression tests were conducted. All tests were loaded up to failure at a constant loading rate of 1 mm/min by using the same MTS loading system as in the Brazilian tests, and the primary failure loads were recorded.

5.3.3 Fracture test of bi-material specimen

As far as the crack propagation study is concerned, same a MTS loading system was used. Then, at a constant rate of 0.5 mm/min, the specimen is led to its maximum strength. The maximum load is recorded. This procedure was applied to all bi-material specimens until failure. And in order to keep in track the failure mode and patterns, photographs of each series of 8 specimens were taken, and exposed in this study.

5.4 Experimental results

This paragraph gathers the results of the experimental investigation carried on all Brazilian discs, cylindrical specimens, and CSTBD specimens. First, the elastic properties of the cement and gypsum are presented. Subsequently, the maximum loads at failure recorded for the bi-material specimens are shown in form of two tables. Then, we use the BEM code to calculate the normalized SIFs,

91

crack initiation angle and propagation path. The input material properties and geometry should obviously be the same as that in the experiments. Finally, the photographs of all bi-material specimens are exposed in order to highlight the crack paths. And then, these experimental observations were compared with numerical predictions.

5.4.1 The strengths of compressive and tensile

(i)　Uniaxial compressive strength (UCS)

A total of 6 uniaxial compression tests were conducted on cylindrical specimens of cement and gypsum in order to determine their UCS. The results of the uniaxial compression tests are reported in Table 6.4.

Table 5.4 cylindrical specimens tested in UCS.

No.		$W_f(KN)$	$D(m)$	$A(m^2)$	UCS (MPa)
Cement	CC-01	90.650	0.045	0.001590	56.997
	CC-02	89.450	0.045	0.001590	56.243
	CC-03	91.020	0.045	0.001590	57.230
Average		90.373	0.045	0.001590	56.823
Gypsum	GC-01	10.250	0.045	0.001590	6.445
	GC-02	10.900	0.045	0.001590	6.853
	GC-03	10.300	0.045	0.001590	6.476
Average		10.483	0.045	0.001590	6.592

Cement　　　　　　　　Gypsum

(ii) Tensile strength

A total of 6 Brazilian tests were conducted on disc of cement and gypsum in order to determine their tensile strength. Table 5.5 gives the thickness t, the failure load W_f, and the tensile strength for each test.

Table 5.5 Brazilian test results on cement and gypsum specimens

No.		$W_f(KN)$	D (m)	t (m)	Tensile strength (MPa)
Cement	CC-01	7.51	0.074	0.01	6.395
	CC-02	7.54	0.074	0.01	6.421
	CC-03	7.56	0.074	0.01	6.438
Average		90.373	0.074	0.01	6.418
Gypsum	GC-01	3.21	0.074	0.01	2.734
	GC-02	3.22	0.074	0.01	2.742
	GC-03	3.24	0.074	0.01	2.759
Average		10.483	0.074	0.01	2.745

Finally, Table 5.6 lists the physical properties of the cement and gypsum specimens including its moisture content, unit weight, porosity, the USC, and tensile strength. According to the classification system recommended by the ISRM 1981, rock may range from extremely weak to extremely strong depending on the uniaxial compressive strength as shown in Table 5.7. Specimens of gypsum and cement, can be summarized as weak rock (R2) and strong rock (R4), respectively.

Table 5.6 Physical properties

	Cement	Gypsum
Water/cement and water/gypsum ratio	0.3	0.5
Water content (%)	3.17	23.42
Density(g/cm^3)	1.999	1.395
Specific gravity, Gs	3.16	2
Void ratio	0.1001	0.4684
Porosity (%)	10	31.898
Setting time (hr)	24	0.5
UCS (MPa)	56.823	6.592
Tensile strength (MPa)	6.418	2.745

Table 5.7 Classification system recommended by the ISRM 1981

Rating	R0	R1	R2	R3	R4	R5	R6
Description of rock strength	extremely weak rock	very weak rock	weak rock	medium strong rock	strong rock	very strong rock	extremely strong rock
UCS (MPa)	≤ 0.98	$0.98 \sim 4.9$	$4.9 \sim 24$	$24 \sim 49$	$49 \sim 98$	$98 \sim 245$	≥ 245

5.4.2 Elastic properties

The determination of elastic properties of the materials which to include the Brazilian disc specimens and the cylindrical specimens both have been suggested by ISRM. The production methods as described below.

Samples CD-01~03, and GD-01~03 have been tested following the procedure described in section 5.3.1. The results obtained for each specimens have been plotted to determine the Young's modulus E and Poisson's ratio v of this isotropic rock. In the Figure 5.11 and 5.12, the Young's module corresponds to the trend line's slope, and the Poisson's ratio is calculated at 50% of the load at failure. The results of Young's modulus, E Poisson's ratio v, and Shear modulus, G are displayed in Table 5.8.

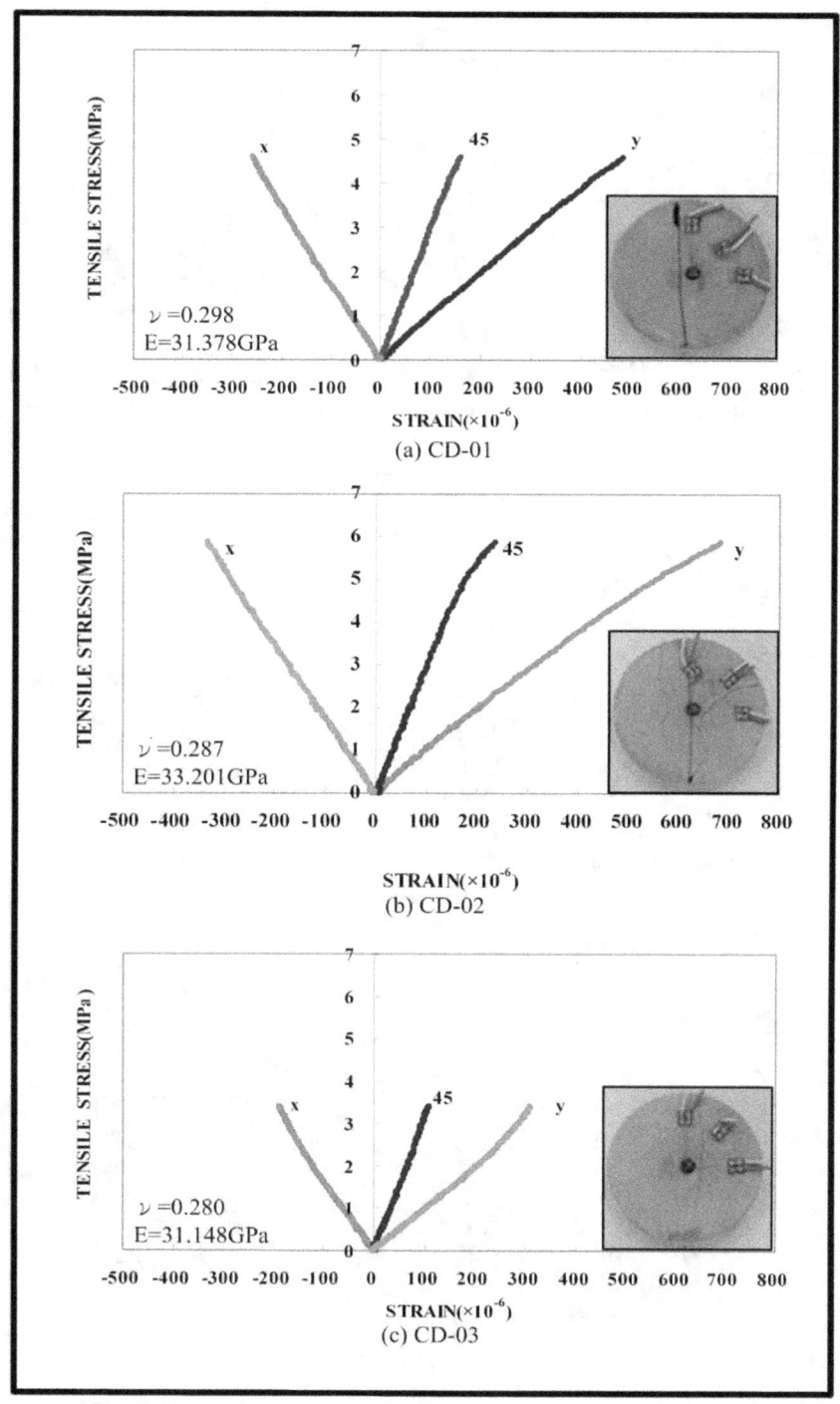

Figure 5.11: Stress versus Strain of cement specimens, determination of *E* and *v*.

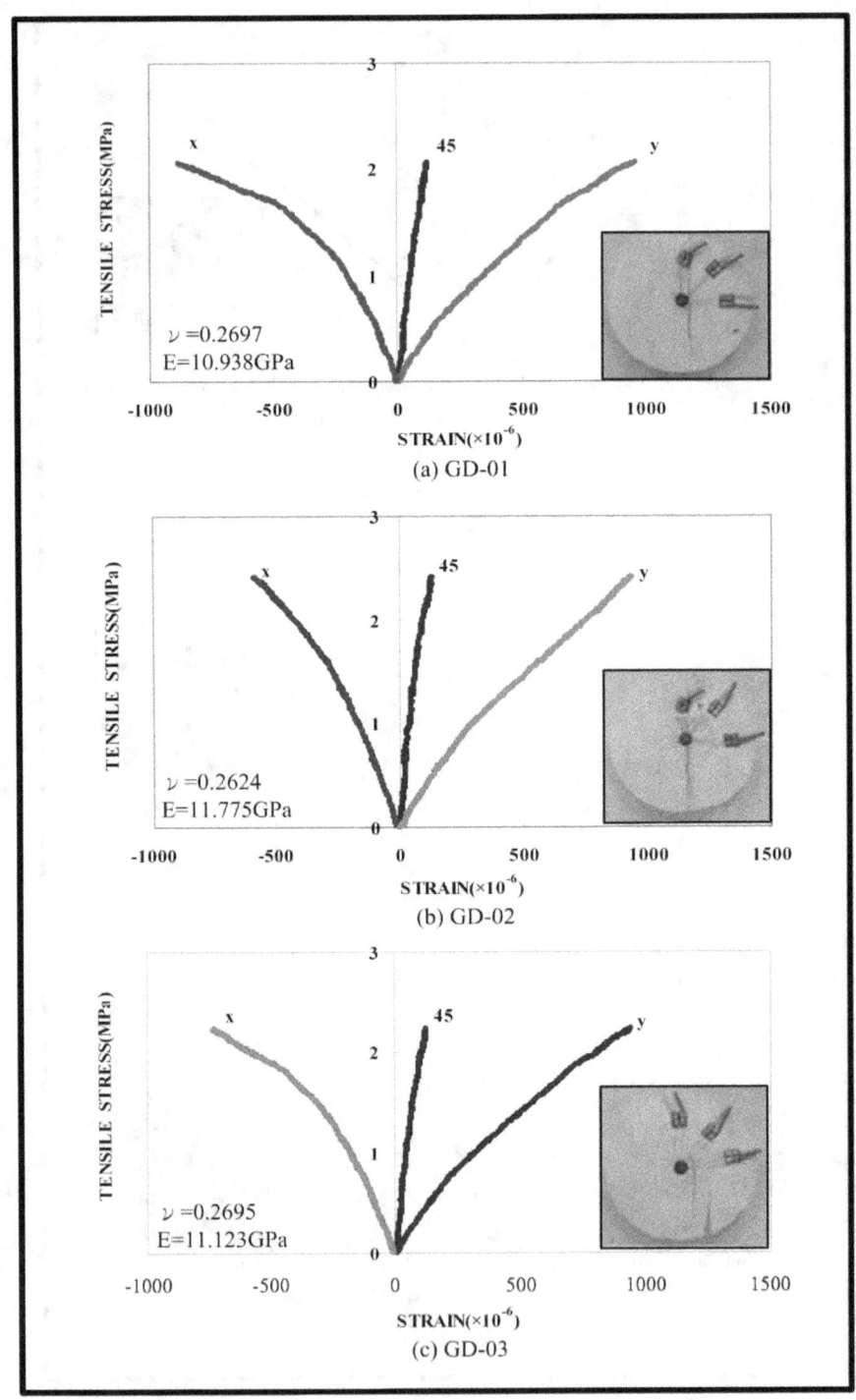

Figure 5.12: Stress versus Strain of cement specimens, determination of E and v.

Figure 5.11(a) shows a linear elastic behavior indicating a constant value of the modulus of elasticity till the point of failure. This type of curve is exhibited by most of the igneous rocks, such as basalt, quartzite, and strong sandstones, etc.

Table 5.8 Elastic constants in the plane of isotropy.

No.		D(mm)	t(mm)	E(GPa)	v	G(GPa)
Cement	CD-01	74	11	31.378	0.299	12.078
	CD-02	74	10.5	33.202	0.287	12.900
	CD-03	74	10.8	31.148	0.280	12.164
	Average			31.909	0.289	12.380
Gypsum	GD-01	74	10.8	10.938	0.270	4.308
	GD-02	74	9.8	11.775	0.262	4.664
	GD-03	74	10.5	11.123	0.269	4.381
	Average			11.279	0.267	4.450

5.4.3 Maximum loads

The maximum load is the load recorded when the first failure plane appears on the surface of specimen. The following Table 5.9 contains the maximum loads recorded for the CSTBD specimens.

Table 5.9 Maximum loads for CSTBD bi-material specimens,

No.	D (mm)	Pre-crack length 2a (mm)	Pre-crack inclined angle β (deg.)	Interface angle λ (deg.)	Normalized SIFs of tip A with the BEM		Normalized SIFs of tip B with the BEM		W_f (KN)
					F_I	F_{II}	F_I	F_{II}	
CGD-1	74	23	0	90	1.765	0.000	0.776	0.000	1.040
CGD-2	74	20.5	45	90	-1.338	-2.857	-1.480	-1.714	1.710
CGD-3	74	23	0	135	1.513	0.000	0.722	0.000	1.053
CGD-4	74	23	45	135	-1.564	-2.560	-1.008	-1.416	1.980
CGD-5	74	23	45	45	-1.257	-1.794	-1.595	-1.830	1.570
CGD-6	74	22.33	0	0	0.965	0.000	0.984	0.000	0.980
CGD-7	74	21.4	45	0	-0.677	-1.345	-2.421	-2.807	1.790

97

5.4.4 Crack initiation and propagation

This paragraph gathers the photographs of all specimens, both sides. For each type of specimen, several parameters have been measured and plotted in tables. These parameters are: (i) d_i : is the distance separating the i tip of the pre-crack from the crack initiation point, propagating direction, (ii) θ_i : is the trigonometric angle between the pre-crack axis and the crack initiation direction, (iii) End Up Points : ξ_i, express the trigonometric angles between the horizontal direction and the locations where the cracks have reached the periphery. (See Figure 5.13)

It is important to notice that the crack path is not necessarily identical in both sides of a given specimen. Therefore, all the values presented in the following tables are an average between the values read on both sides. Moreover, in most cases, after first cracks occurred, the loading in process caused secondary fractures. These fractures are essentially due to the closure of the pre-crack; therefore cracks initiate from the periphery of the disc by tension, and propagate toward the pre-crack tip. In this study, the secondary fractures are not taken into consideration.

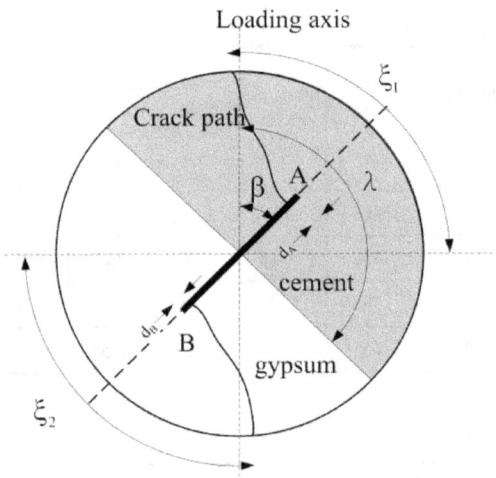

Figure 5.13: Measuring d_i, and End-UP Points ξ_i.

98

Based on the calculation of the SIFs and crack initiation angle for each increment, the procedure of fracture propagation can be simulated. Now, we focus on the simulation of fracture propagation path with the BEM program to illustrate the incremental length of crack extension. Because the present fracture propagation analysis regards the growth of a single crack, the incremental length of the crack extension can be defined arbitrarily (see section 3.4.2). However, some restrictions on the length are introduced to ensure efficiency of the numerical analysis. For the sake of simplicity, the incremental length of the crack extension is discretized by the fixed length of new constant boundary elements. In order to avoid numerical problems concerned with the relative length of neighboring elements, the incremental length of the crack extension is kept constant, between convenient limiting bounds defined in terms of the length of the initial crack tip element (Portela, 1993).

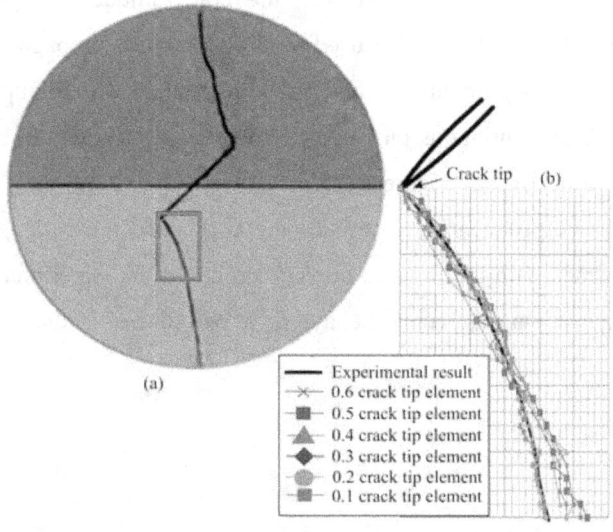

Figure 5.14: Comparison of experimental and numerical crack propagation path (lower tip only).

In order to define the incremental length of the crack extension; we take the

99

six different lengths of incremental elements to simulate the crack propagation path. Then, the numerical results are checked on CGD2-01 specimen as shown in Figure 5.14(a). The Figure 5.14(b) shows the variation of crack propagation paths for different incremental lengths for cement/gypsum specimen of the local crack path determined numerically and experimentally. A good agreement is found between the experimental results of CGD2-01 specimen and the numerical results that take the incremental length are equal to 0.2 time of crack tip element length. Therefore, we can be defining the crack increment length is fixed at 0.2 time of crack tip element length in this simulation.

Photographs showing specimens after failure and the crack propagation paths is given in appendix A, respectively. Table 5.8 gives a comparison between the crack initiation angle θi measured experimentally and those predicted numerically with the BEM code. It can be seen that our numerical procedure based on the BEM can predict well the crack initiation angle in CSTBD experiment. The BEM was also used to simulate crack propagation in CSTBD specimens. The outer boundary and crack surface of CSTBD specimens were meshed with 28 continuous quadratic elements and 10 discontinuous quadratic elements, respectively. Figure 5.15(a) and 5.16(a) show the observed and predicted crack propagation paths of CGD1-01 and CGD2-01 specimens. The numerical crack path is shown to be very similar to the experimental crack path in those figures. On the fracture behavior, we will discuss in the next section.

Table 5.8 Comparison between crack initiation angles predicted numerically and observed experimentally in the test on CSTBD bi-material specimens.

No.	Pre-crack length $2a$ (mm)	Pre-crack inclined angle β (deg.)	Interface angle λ (deg.)	$\theta_{0(A)}$			$\theta_{0(B)}$		
				test	Num.	Diff. (%)	test	Num.	Diff. (%)
CGD1	23	0	90	0	0	0	0	0	0
CGD2	20.5	45	90	77.33	78.517	-1.53	87	87.378	-0.43
CGD3	23	0	135	0	0	0	0	0	0
CGD4	23	45	135	59.000	59.425	-0.72	59.330	57.731	2.69
CGD5	23	45	45	80.330	81.370	-1.29	85.670	88.230	-2.99
CGD6	22.33	0	0	0	0	0	0	0	0
CGD7	21.4	45	0	82.500	81.570	1.12	69.500	67.830	2.4

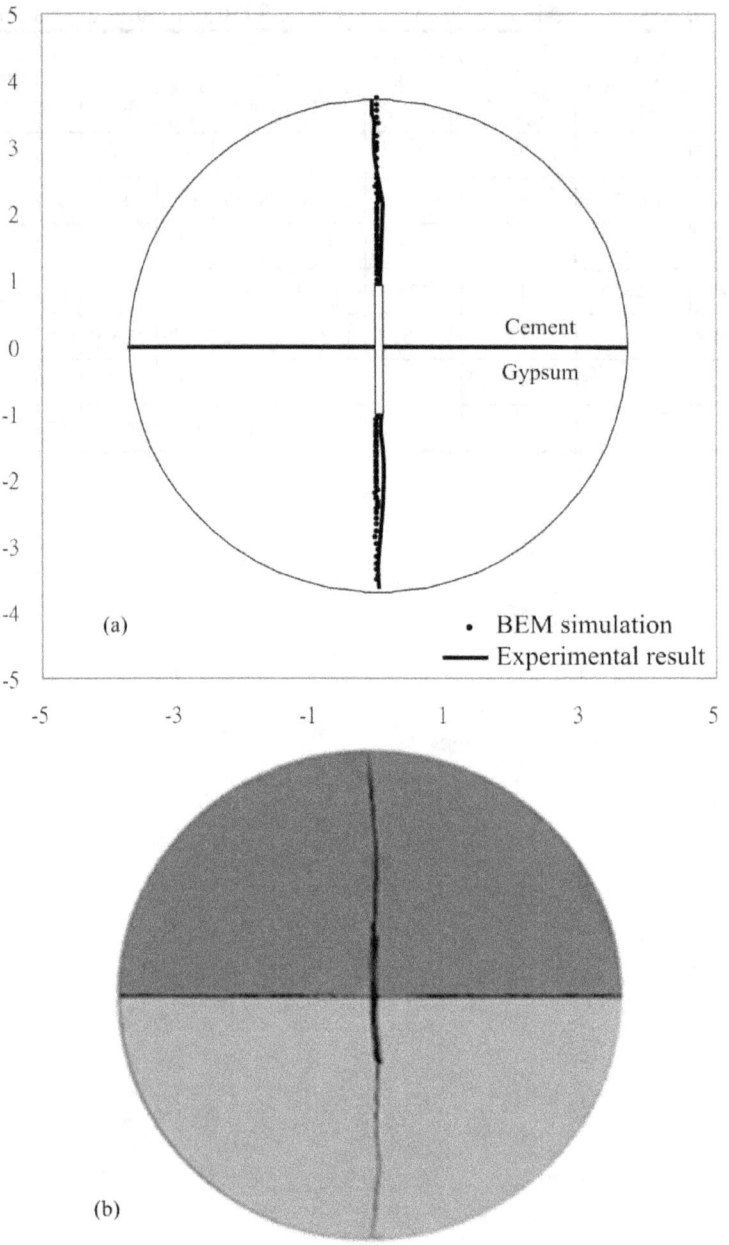

(a)

Cement

Gypsum

• BEM simulation
— Experimental result

(b)

Figure 5.15: Comparison between experimental observations and numerical predictions for specimen CGD1-01: (a) propagation of a CSTBD specimen with $\lambda = 90°$ and $\beta = 0°$ and (b) photograph of specimen CGD1-01 after failure.

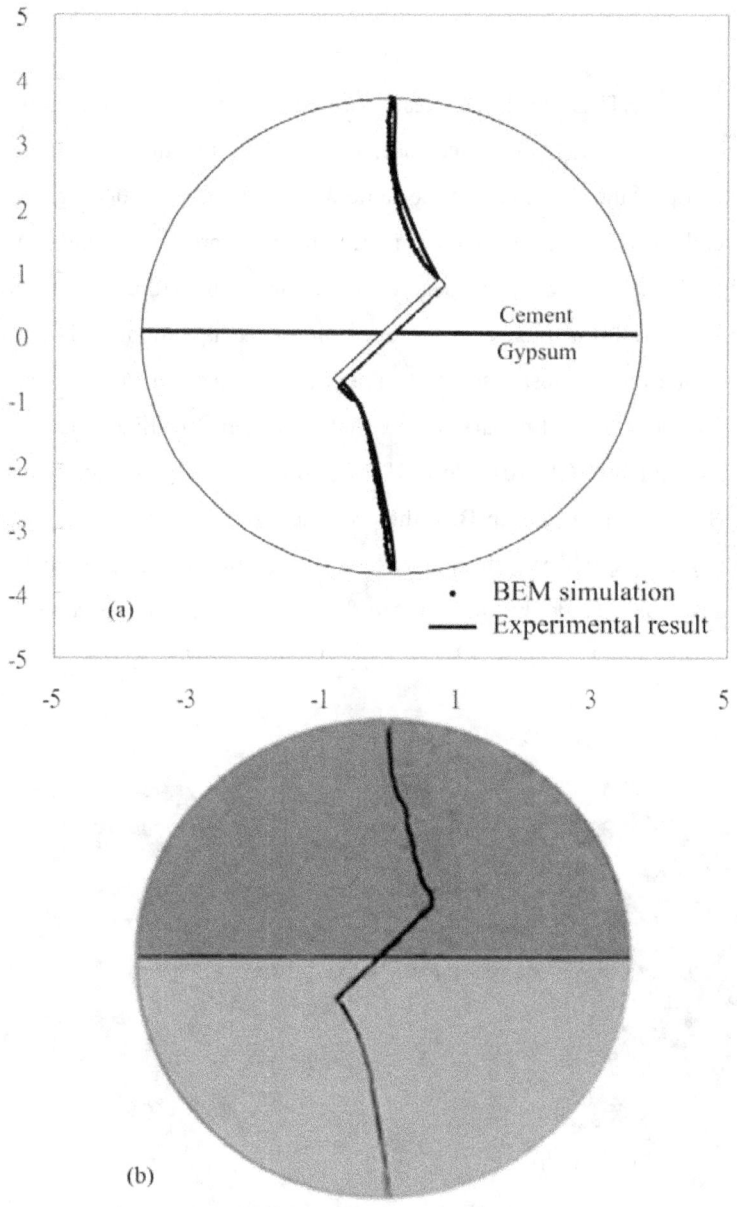

Figure 5.16: Comparison between experimental observations and numerical
predictions for specimen CGD2-01: (a) propagation of a CSTBD specimen with
$\lambda = 90°$ and $\beta = 45°$ and (b) photograph of specimen CGD2-01 after failure.

5.5 Interpretation of Experimental Results

In the case of CSTBD specimens, test revealed that the crack initiates first in the central pre-crack, and as β decrease, d_i decrease, i.e. the crack tends to initiate at the tip of the pre-crack. In the same way, as β decrease, θ_i decrease, i.e., the crack tends to initiate in the same direction than the pre-crack. One can also observe that the crack always propagate in the direction of the loading zones where the stress field in intense. Also, once the crack has initiated, it changes direction to take the shortest distance to reach the loading zones. Furthermore, we notice that pure mode I occurs for β equal to 0°. One peculiarity is to point out for the specimens CGD5-01 and CGD7-01: for $\beta = 45°$, two crack paths are observed. Both initiate at the tip B of the pre-crack, but one propagates directly to the loading zones, whereas the other starts in direction of the pre-crack axis (interface axis), and suddenly kinks in the direction of the loading zones. (See Figure 5.17)

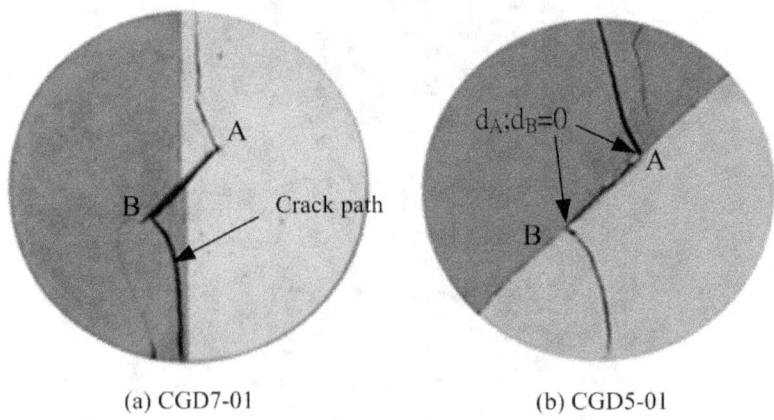

(a) CGD7-01   (b) CGD5-01

Figure 5.17: Double fracture paths for $\beta = 45°$

5.6 Numerical analysis of a CSTBD specimen

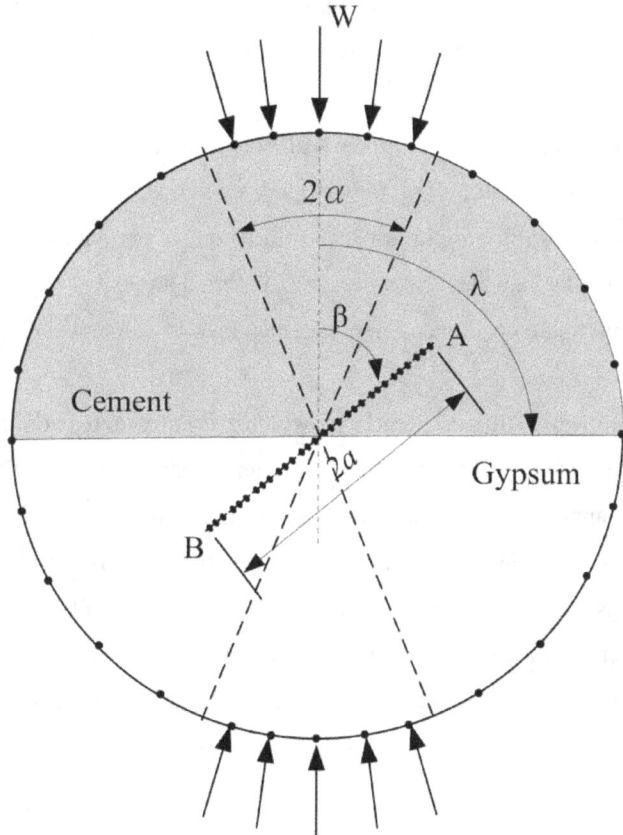

Figure 5.18: Geometry of a Cracked Straight Through Brazilian Disc (CSTBD) specimen of cement/gypsum bi-materials under segmental loading.

A 74 mm diameter CSTBD loaded by a segmental force W ($2\alpha = 10°$) is now considered as shown in Figure 5.18. The elastic constants of CSTBD specimen were given by Table 5.8. The outer boundary and crack surface of CSTBD specimens were meshed with 28 continuous quadratic elements and 10 discontinuous quadratic elements, respectively. The variation of the SIFs and initial angle θ with the crack inclined angle β which varies from 0° to 180° with increments of 15°, respectively. The plane stress conditions were supposed. Here the interface orientation λ is fixed at 90° and crack length is such that $2a = 20$ mm. Calculations were carried out by our BEM code. Figure 5.19 and Figure 5.20 represents the variations of normalized SIFs (F_I and F_{II}) and the initial angle θ according to the inclined angle β, respectively.

For Figure 5.19 shows that mode I SIFs reaches a positive maximum value when the crack is parallel to the applied load (i.e., $\beta = 0°$) and decreases smoothly to a negative minimum when the crack is perpendicular to the load. The mode II SIFs vanishes when the crack inclined angle $\beta = 0°$ and $\beta = 90°$ and reaches a maximum absolute value when the crack is oriented at 45° (as tip B).

The initial angle (θ) is the angle by which the crack extension deviates from the direction of the original tip. Numerical result are displayed with the theoretical analysis of crack propagation under mixed mode I–II loading condition based on the maximum tensile stress (σ-criterion). The crack initiation is more sensitive to the degree of crack sharpness at small crack orientations with respect to the loading zone. At higher crack orientations ($\beta > 30°$), the degree of crack sharpness has not significant effect on the initial angle.

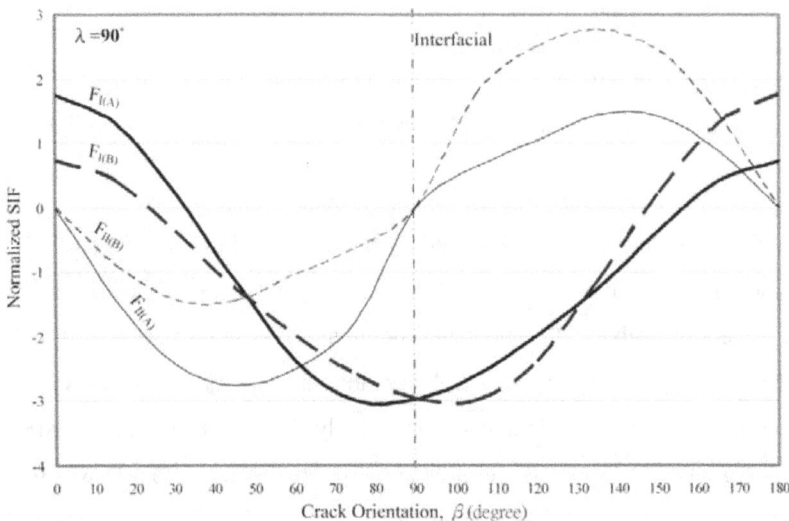

Figure 6.34: Variation of F_I and F_{II} according to the crack inclined angle β.

106

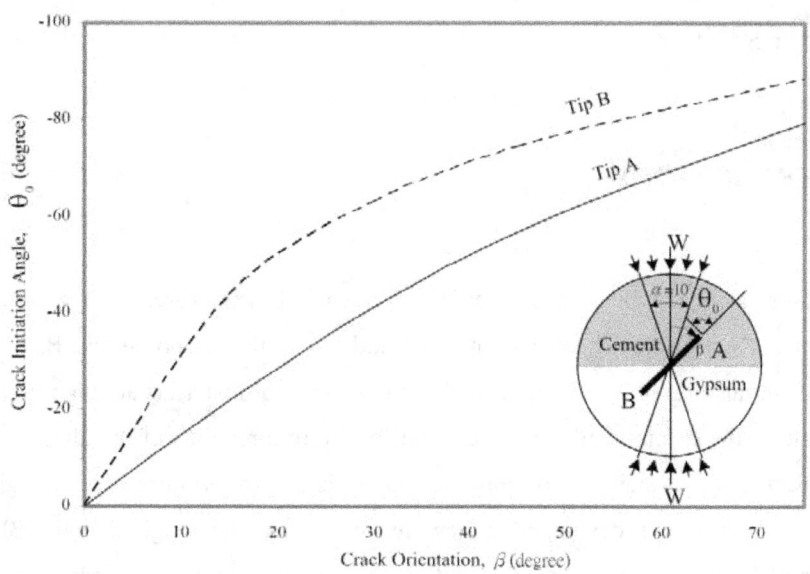

Figure 6.35: Variation of the crack initial angle θ with the crack inclined

angle β.

Chapter 6

Conclusions

This chapter summarizes the key achievements of the research work presented in the disquisition. The values and limitations of the proposed the BEM of bi-materials in numerically solving the problems of bi-materials are highlighted. Some of the recommendations are made for future directions of this thesis with respect to the bi-materials of numerical analysis and experimental technologies. In this study, we developed a new technique for the single-domain BEM formulation in which neither the artificial boundary nor the discretization along the un-cracked interface is necessary. This single-domain BEM formulation was presented by Chen et al., 1996 for homogeneous materials. Now, we combine it with the Green's functions of bi-materials (Pan and Amadei, 1999) can be extended to anisotropic bi-materials. The major achievements of the research work are summarized as follows.

A decoupling technique can be used to determine the SIFs of the mixed mode and the oscillation on interfacial crack, based on the relative displacements at the crack tip. Five types of three-node quadratic elements are utilized to approximate the crack tip as well as the outer boundary; a crack surface and an interfacial crack surface are evaluated using the asymptotical relation between the SIFs. Since the interfacial crack has an oscillation singular behavior, we take a special crack-tip element (Gao et al., 1992) to exactly capture this behavior.

Calculation of the SIFs is conducted for several situations, like cracks along or off an interface. Numerical results show that the proposed method is very accurate even with relatively coarse mesh discretization. In addition, the proposed BEM program is not only very accurate but also very fast. The calculation time for each example is typically 30s by BEM code with a Personal Computer, PC (Intel Pentium 4, 2.4 GHz CPU, 1GB RAM).

This study presents the development of BEM procedure based on the maximum circumferential stress criterion was used to predict the crack initiation direction and the crack propagation path in bi-materials under mixed mode loading. An agreement was found between crack initiation and propagation predicted with the numerical result and experimental result observations reported by previous researchers of isotropic materials. And numerical simulations of crack initiation and propagation in bi-material specimens of CSTBD were also found to be very similar to the experimental results.

The Brazilian test by laboratory testing has been applied to obtain the elastic constants of isotropic materials and their indirect tensile strength. A Brazilian disc of cement and gypsum are loaded across the diameter until it splits. Through the maximum failure loading and stress-strain curve, the independent constants have been obtained in the experiment. A total of 6 Brazilian tests were conducted on disc of cement and gypsum in order to determine their tensile strength. Mixed tensile splitting and shear was also observed in certain cases.

In CSTBD of crack initiation and propagation, in the most general case, tests revealed that the crack initiates first in the middle of the pre-crack, and as crack angle β decreases, the crack tends to initiate at the tip of the pre-crack. In the same way, as β decreases, the crack tends to initiate in the same direction than the pre-crack. It has also been observed that the crack always propagates toward the loading zones where the stress field is intense. Also, once the crack has initiated, it changes direction to take the shortest path to reach the loading zones. One interesting point has attracted attention: on interfacial crack of CSTBD for $\beta = 45°$, two directions of crack occur at the same tip. Both initiate at the tip of the pre-crack, but one propagates directly toward the loading zones, whereas the other starts in direction of the pre-crack axis, and suddenly kinks toward the loading zones.

Bibliography

Aliabadi, M. H., *Fracture of Rock*, WITpress, Boston, (1999).

Aliabadi, M. H., Boundary element formulation in fracture mechanics, *Applied Mechanics Review*, *ASME.*, Vol. 50, pp. 83-96, (1997).

Al-Shayea, N. A., Crack propagation trajectories for rocks under mixed mode I-II fracture, *Engineering Geology*, Vol. 81, no. 1, pp. 84–97, (2005).

Al-Shayea, N. A., Khan, K. and Abduljauwad, S.N. Effects of confining pressure and temperature on mixed mode (I-II) fracture toughness of a limestone rock, *Int. J. Rock Mech. Min. Sci.*, Vol. 37, pp. 629-643, (2000).

Amadei, B., Importance of anisotropy when estimating and measuring in-situ stresses in rock, *Int. J. Rock Mech. Min. Sci. Geom. Abs.*, Vol. 33, no.3, pp. 293-325, (1996).

Amadei, B. and Pan, E., Gravitational stresses in anisotropic rock masses with inclined strata, *Int. J. Rock Mech. Min. Sci. Geom. Abs.*, Vol. 29, no. 3, pp. 225-236, (1992).

Amadei, B., Savage, W.Z. and Swolfs, H.S. Gravitational stresses in anisotropic rock masses, *Int. J. Rock Mech. Min Sci. Geom. Abs.*, Vol. 24, pp. 5-14, (1987).

Amadei, B. and Savage, W. Z., *Gravitational stresses in regularly jointed rock masses – A keynote lecture,* Proceed. of the Int. Symp. on Fundamentals of Rock Joints, Bjorkliden, O. Stephanson (Ed.), Centek Publishers, Lulëa, Sweden, (1985).

Amadei, B., *Rock anisotropy and the theory of stress measurements*, Springer-Verlag, New York, (1983).

Atkinson, C. and Craster, R. V., Theoretical aspects of fracture mechanics, *Prog. Aerospace Sci.*, Vol. 31, pp. 1-83, (1995).

Atkinson, B. K., *Fracture mechanics of rocks : experimental fracture mechanics data of rock and minerals.*, Academic press, London, (1987).

Atkinson, C., The interface crack with a contact zone (the crack of finite length), *Int. J. Fracture,* Vol. 19, pp. 131-138, (1983).

Atkinson, C., The interface crack with a contact Zone (an analytical treatment), *Int. J. Fracture.*, Vol. 18, pp. 161-177, (1982).

Atkinson, C., Smelser, R. E. and Sanchez, J., Combined mode fracture via the cracked Brazilian disk test, *Int. J. Fracture*, Vol. 18, No. 4, pp. 279-291, (1982).

Atkinson, C., On stress singularities and interfaces in linear fracture mechanics, *Int. J. Fracture*, Vol. 13, pp. 807-820, (1977)

Atkinson, C., *Dynamic crack problems in dissimilar media, mechanics of fracture IV: elastodynamic crack problems,* Edited by G.C. Shi, Noordhoff, Leyden, (1977).

Awaji, H. and Sato, S., Combined mode fracture toughness measurement by the disk test, *J. Eng. Mater. Tech. Trans. ASME*, Vol. 100, No. 4, pp. 175-182, (1978).

Bandis, S. C., Lumsden, A. C. and Barton, N. R., Fundamentals of rock joint deformation, *Int. J. Rock Mech. Min. Sci. Geom. Abs.*, Vol. 20, no.6, pp. 249-268, (1983).

Barsoum, R.S., On the use of isoparametric finite elements in LEFM, *Int. J. Num. Meth. in Eng.*, Vol. 10, pp. 25-37, (1976).

Becker, A. A., *The boundary element method in engineering*, McGraw-Hill Book Company, UK, (1992).

Bhattacharya, J. P. and Walker, R. G., Fancies models response to sea level change, *Geol. Assoc. Canada.*, pp. 157-178, (1992).

Bieniawski, Z. T. and Hawkes, I., Suggested methods for determining tensile strength of rock materials. *Int. J. Rock Mech. Min. Sci. & Geomech. Abstr.*, Vol. 15, pp. 99-103, (1978).

Blandford, G. E., Ingraffea, A. R. and Liggett, J. A., Two-dimensional stress intensity factor computations using the boundary element method, *Int. J. Num. Meth. in Eng.*, Vol. 17, pp. 387-404, (1981).

Bobet, A., Modelling of crack initiation, propagation and coalescence in uniaxial compression. *Rock Mechanics and Rock Engineering,* Vol. 33, no. 2, pp. 119-139, (2000).

Bogy, D. B. and Wang, K. C., Stress singularities at interface comers in bonded dissimilar isotropic elastic materials, *Int. J. Solids Stru.*, Vol. 7, pp. 993-1005, (1971).

Bogy, D. B., On the plane elastostatic problems of a loaded crack terminating at a material interface,

J. Appl. Mech., Vol. 38, pp. 911-918, (1971).

Bogy, D. B., Edge-bonded dissimilar orthogonal elastic wedge under normal and shear loading, *J. Appl. Mech.,* Vol. 35, pp. 460-466, (1968).

Bowen, J. M. and Knauss, W. G., An experimental study of interfacial crack kinking, *Experimental Mech.,* Vol. 3, pp. 37-43, (1993).

Brady, B. H. G., Lemos, J. V. and Cundall, P. A., *Stress measurements schemes for jointed and fractured rock*, Proceed. of the Int. Symp. on Rock Stress and Rock Stress Measurements, Stockholm, O. Stephanson (Ed.), Centek Publishers, Lulëa, Sweden, (1986).

Bray, J. W., A student of jointed and fractured rock, Fracture patterns and their failure characteristics, *Rock Mechanics and Engineering Geology*, Vol. 5, Springer, Berlin, pp. 117-136, (1967).

Brebbia, C. A., and Dominguez, J., *Boundary element an introductory course*, Second Edition, Computational Mechanics Publications, Southampton, (1992).

Brown, E. T., *Rock characterization testing and monitoring : ISRM suggested methods*, Pergamon press, Oxford, (1981).

Chen, C. S., Ke, C. C., and Tu, C. H., Evaluating the stress intensity factors of anisotropic bimaterials using boundary element method, *International Journal for Numerical and Analytical Methods in Geomechanics*, Vol.32, no.11, pp. 1341-1364, (2008).

Chen, C. S., Pan, E. and Amadei, B. Determination of deformability and tensile strength of anisotropic rock using Brazilian tests, *Int. J. Rock Mech. Min. Sci.,* Vol. 35, no. 1, pp. 43-61, (1998).

Chen, C. S., *Characterization of deformability, strength, and fracturing of anisotropic rocks using Brazilian tests*, Ph.D. Thesis, Department of Civil, Environmental and Architectural Engineering, University of Colorado, (1996).

Chen, J. T. and Hong, H. K., Review of dual boundary element methods with emphasis on hypersingular integrals and divergent series, *Applied Mechanics Review, ASME.,* Vol. 52, no.1, pp. 17-33, (1999).

Chen, J. T. and Wang, W. C., Experimental analysis of an arbitrarily inclined semi-infinite Crack Terminating at the Bi-material Interface, *Experimental Mech.,* Vol.24, pp. 7-16, (1997).

Cho, S. B., Lee, K. B., Choy, Y. S. and Yuuki, R., Determination of stress intensity factors and boundary element analysis for interface cracks in dissimilar anisotropic materials, *Eng. Fracture Mech.,* Vol. 43, no. 4, pp. 603-614, (1992).

Cisilino, A.P., Aliabadi, M.H., Three-dimensional boundary element analysis of fatigue crack growth in linear and non-linear fracture problems, *Eng. Fracture Mech.,* Vol. 63, pp. 713-733, (1999).

Comninou, M., An Overview of interface cracks. *Eng. Fracture Mech.,* Vol. 37, 197-208, (1990).

Comninou, M. and Dundurs, J., On the behavior of interface cracks, *Res. Mechanics,* Vol. 1, pp. 246-264,(1980).

Comninou, M. and Schmueser, D., The interface crack in a combined tension compression and shear field, *J. Appl. Mech.,* Vol. 46, pp. 345-348, (1979).

Comninou, M., The interface crack in a shear field, *J. Appl. Mech.,* Vol. 45, pp. 287-290, (1978).

Comninou, M., The interface crack. *J. Appl. Mech.,* Vol. 44, pp. 631-636, (1977)

Cook, T. S. and Erdogan, F., Stresses in bonded materials with a crack perpendicular to the interface. *Int. J. Engng Sci.,* Vol. 10, pp. 677-697, (1972).

Crouch, S. L. and Starfield, A. M., *Boundary element methods in solid mechanics*, George Allen and Unwin Publishers, London, (1983).

Cruse, TA., *Boundary element analysis in computational fracture mechanics*, Dordrecht: Kluwer Academic, (1988).

Cundall, P. A., A computer model for simulating progressive large scale movements in Blocy rock systems, *ISRM Symp., Nancy, France, Proc 2.,* pp. 129-136, (1971).

Dale, F. and Erdogan, F., On the mechanical modeling of the interfacial region in bonded half-planes, *J. Appl. Mech.,* Vol. 55, pp. 317-324, (1988).

Dong, S., Wang, Y. and Xia, Y., Stress intensity factors for central cracked circular disk subjected to compression, *Engineering Fracture Mechanics.,* Vol. 71, pp. 1155-1168, (2004).

Dundurs, J., Discussion on "Edge-bonded dissimilar orthogonal elastic wedges under normal and shear loading"., *J. Appl. Mech.,* Vol. 30, pp. 650-652, (1969).

Dwyer, J. F. and Pan, E. Edge function analysis of stress intensity factors in cracked anisotropic plates. *Int. J. Fracture,* Vol. 72, pp. 327–342, (1995).

Elliott, T., *Sedimentary environments and facies*, Deltas, in Reading, H. G., ed., (second edition): Oxford, Blackwell Scientific, (1986).

111

England, A. H., A crack between dissimilar media, *Trans. ASME, J. Appl. Mech.*, Vol. 32, pp. 400-402, (1965).

Erdogan, F., Kaya, A. C. and Joseph, P. F., The crack problem in bonded non-homogeneous materials, *J. Appl. Mech.*, Vol. 58, pp. 410-416, (1991).

Erdogan, F. and Biricikoglu, V., Two bonded half plane with a crack going through the interface, *Int. J. Eng. Sci.*, Vol. 11, pp. 745-766, (1973).

Erdogan, F., Gupta, G. D. and Cook, T. S., Numerical solution of singular integral equations, *Mechanics of Fracture I: Methods of Analysis and Solutions of Crack Problems*, Edited by G.C. Sih, Noordhoff, Leyden, (1973).

Erdogan, F., Stress distribution in a non-homogeneous elastic plane with cracks, *J. Appl. Mech.*, Vol. 30, pp. 232-237, (1963).

Erdogan, F. and Sih, G. C., On the crack extension in plates under loading and transverse shear. *J. Basic Eng.*, Vol. 85, pp. 519-527, (1963).

Fischer M. P., Elsworth, D., Alley, R. B. and Engelder, T., Finite element analysis of the modified ring test for determining mode I fracture toughness, *Int. J. Rock Mech. Min. Sci. & Geomech. Abstr.*, Vol. 33, no. 1, pp. 1-15, (1996).

Fowell, R. J. and Xu, C., The use of the cracked Brazilian disc geometry for rock fracture investigations. *Int. J. Rock Mech. Min. Sci. & Geomech. Abstr.*, Vol. 31, no. 6, pp. 571-579, (1994).

Gandhi, K. R., Analysis of inclined crack centrally placed in an orthotropic rectangular plate. *J. Strain analysis*, Vol. 7, no. 3, pp. 157-162, (1972).

Gao, H., Abbudi, M. and Barnett D. M., Interfacial crack-tip field in anisotropic elastic solids. *J. Mech. Phys. Solids*, Vol. 40, pp. 393-416, (1992).

Gautesen, A. K. and Dundurs, J., The interface crack under combined loading, *J. Appl. Mech.*, Vol. 55, pp. 580-586, (1988).

Gautesen, A. K. and Dundurs, J., The interface crack in a tension field, *J. Appl. Mech.*, Vol. 54, pp. 93-98, (1987).

Gdoutos, E. E., Photoelastic analysis of stress field around singular points or rigid inclusions, *J. Appl. Mech.*, Vol. 49, pp. 236-239, (1982).

Germanovich, L. N. and Dyskin, A. V., Fracture mechanisms and instability of openings in compression, *Int. J. Rock Mech. Min. Sci.*, Vol. 37, pp. 263-284, (2000).

Germanovich, L. N., Salganik, R. L., Dyskin, A. V. and Lee, K. K., Mechanisms of brittle fracture of rocks with multiple pre-existing cracks in compression, *Pure and Applied Geophysics*, Vol. 143(1/2/3), pp. 117-149, (1994).

Geubelle, P. H. and Knauss, W. G., Crack propagation at and near bimaterial interfaces: linear analysis, *J. Appl. Mech.* Vol. 61, pp. 560-566, (1994).

Goodman R. E. *Methods of geological engineering*, West Publishing Company, St. Paul, Minn., (1976).

Gramberg, J., *A non-conventional view on rock mechanics and fracture mechanics*, Published of the Commission of the European Communities by A. A. Balkema, Rotterdam, (1989).

Griffith, A. A., The phenomena of ruptures and flow in solids, *Phil. Trans. Roy. Soc. London*, Series A, Vol. 221, pp. 163-198, (1920).

Guo, H., Aziz, N. I. and Schmidt, L. C., Rock fracture-toughness determination by the Brazilian test, *Engineering Geology*, Vol. 33, pp. 177-188, (1993).

Hallbauer, D. K., Wagner, H. and Cook, N. G. W., Some observations concerning themicroscopic and mechanical behavior of quartzite specimens in stiff triaxial compression tests, *Int. J. Rock Mech. Min. Sci.*, Vol. 10, pp. 713-726, (1973).

He, M. Y. and Hutchinson, J. W., Kinking of a crack out of an interface, *J. Appl. Mech.*, Vol. 56, pp. 270-278, (1989).

Henshell, R. D. and Shaw, K. G., Crack tip elements are unnecessary, *Int. J. Arum. Meth. in Eng.*, Vol. 9, pp. 495-507, (1975).

Hoek, E. and Bieniawski, Z. T., Brittle fracture propagation in rock under compression, *Int. J. Fracture*, Vol. 1, pp. 137-155, (1965).

Hondros, G., The evaluation of Poisson's ratio and the modulus of materials of a low tensile resistance by the Brazilian (indirect tensile) test with particular reference to concrete. *Aust. J. App. Sci.*, Vol. 10, pp. 428-434, (1959).

Huang, C. Y. and Cheng, Y. M., Oligocene and Miocene planktic foraminiferal biostratigraphy of northern Taiwan, *Proc. Geol. Soc. China*, Vol. 26, pp. 21-56, (1983).

Hutchinson, J. W., Mear, M. and Rice, J. R., Cracks paralleling an interface between dissimilar materials, *J. Appl. Mech.*, Vol. 54, pp. 828-832, (1987).

Inglis, C., Stresses in a plate due to the presence of cracks and sharp corners, *Trans. Inst. Naval Architects*, Vol. 55, pp. 219-241, (1913).

International Society for Rock Mechanics (ISRM), *Rock Characterization, Testing and Monitoring, ISRM Suggested Methods*, Brown, E. T. (Editor). Pergamon Press, Oxford, (1981).

Irwin, G. R., Analysis of stresses and strains near the end of a crack, *Trans. ASME, J. Appl. Mech.*, Vol. 24, pp. 361-364, (1957).

Isida, M. and Noguchi, H., Formulae of stress intensity factors of branched cracks in plane problems, *Trans. Japan Soc. Mech. Engrs.*, Vol. 49, No. 440, pp. 469-479, (1983).

Isida, M. and Noguchi, H., Plane elastostatic problems of bonded dissimilar materials with an interface crack and arbitrarily located cracks, *Trans. Japan Soc. Mech. Engrs.*, Vol. 49, pp. 137-146, (1983).

Isida, M. and Noguchi, H., Plane problems of arbitrarily oriented cracks in bonded dissimilar materials, *Trans. Japan Soc. Mech. Engrs.*, Vol. 49, pp. 36-45, (1983).

Isida, M., Tension of a half plane containing array cracks, branched cracks and cracks emanating from sharp notches, *Trans. Japan Soc. Mech. Engrs.*, Vol. 45, no. 392, pp. 306-317, (1979).

Isida, M., Arbitrary symmetric loading problems of centrally cracked rectangular plates, *Trans. Japan Soc. Mech. Engrs.*, Vol. 42, no. 362, pp. 3019-3030, (1976).

Isida, M., Arbitrary loading problems of doubly symmetric regions containing a central crack, *Engineering Fracture Mechanics*, Vol. 7, pp. 505-514, (1975).

Jing, L., A review of techniques, advances and outstanding issues in numerical modelling for rock mechanics and rock engineering, *Int. J. Rock Mech. Min. Sci.*, Vol. 40, pp.283-353, (2003).

John, K. W., *Civil engineering approach to evaluate strength and deformability of regular jointed rock. Rock Mechanics – Theory and Practice*, W.H Somerton (Ed.), AIME, New York, (1970).

Kirsch, G., Die theorie der elastizitat und die bedurfnisse der festigkeitslehre, *Veit. Ver. Deut. Ing.*, Vol. 42, pp. 797-807, (1898).

Ku, C. Y., *Modeling of jointed rock masses based on the numerical manifold method*, Ph. D. Dissertation, Department of Civil and Environmental Engineering, University of Pittsburgh, (2001).

Kundu, T., Transient response of an interface-crack in a layered plate, *J. Appl. Mech.* Vol. 53, pp. 579-586, (1985).

Kuo, A. Y., Transient stress intensity factors of an interfacial crack between two dissimilar anisotropic half-spaces, part I: orthotropic materials. *J. Appl. Mech.*, Vol. 51, pp. 71-76, (1984).

Kwon, Y.W. and Dutton, R., Boundary element analysis of cracks normal to bimaterial interfaces, *Eng. Fracture Mech.*, Vol. 40, pp. 487-491, (1991).

Lekhnitskii, S. G. *Theory of elasticity of an anisotropic elastic body*, translated by P. Fern, Holden-Day, San Francisco, (1963).

Lekhnitskii, S. G., *Anisotropic plates*, translated by S.W. Tsai, Gordon and Breath, (1957).

Lempriere, B.M., Poisson's ratios in orthotropic materials, *J. American Institute Aeronaut Astronaut*,Vol. 6, pp. 2226-2227, (1968).

Liao, J. J., *Stress distribution in anisotropic rock masses, ridges and valleys*. Ph.D. Thesis, Department of Civil, Environmental and Architectural Engineering, University of Colorado, (1990).

Libatskii, L. L. and Kovchik, S. E., Fracture of discs containing cracks, *Soviet Materials Science*, Vol. 3, pp. 334-339, (1967).

Liechti, K. M. and Knauss, W. G., Crack propagation at material interfaces: experiments on mode interaction, *Experimental Mech.*, Vol. 22, pp. 383-391, (1982).

Lim, I. L. and Johnston, I. W., Stress intensity factors for semi-circular specimens under three-point bending, *Eng. Fracture Mech.*, Vol. 44, no. 3, pp. 363-382, (1993).

Lin, K.Y. and Mar, J. W., Finite element analysis of stress intensity factors for cracks at a bi-material interface, *Int. J. Fracture*, Vol. 12, pp. 521-531, (1976).

Lu, H. and Chiang, F. P., Photo elastic determination of stress intensity factor of an interfacial crack in a bi-material, *J. Appl. Mech.*, Vol. 60, pp. 93-l00, (1993).

Lu, M. and Erdogan, F., Stress intensity factors in two bonded layers containing cracks perpendicular to and on the interface - I: analysis, *Eng. Fracture Mech.*, Vol. 18, pp. 491 -506, (1983).

Lu, M. and Erdogan, F., Stress intensity factors in two bonded layers containing cracks perpendicular to and on the interface - II: solution and results, *Eng. Fracture Mech.*, Vol. 18, pp. 507-528, (1983).

Malyshev, B. M. and Salganik, R. L., The strength of adhesive joints using the theory fracture, *Int. J. Fracture*, Vol. 1, pp. 114-128, (1965).

Meguid, S. A., Tan, M. A. and Zhu, Z. H., Analysis of cracks perpendicular to bimaterial interfaces Using a Novel Finite Element, *Int. J. Fracture.*, Vol. 73, pp. 1-23, (1995).

Meguid, S.A., *Engineering fracture mechanics*, Elsevier Applied Science (1989).

Miskioglu, I., Villmann, C. R., Pawloski, J. S. and Pariseau, D. M., A photoelastic and FEM analysis of interfacial crack propagation, *Experimental Mech.*, Vol. 31, pp. 135- 139, (1991).

Murakami, Y., A simple procedure for the accurate determination of stress intensity factors by finite element method, *Engineering Fracture Mechanics.*, Vol. 8, pp. 643-655, (1976).

Pageau, S. S., Gadi, K. S., Biggers, S. B. and Joseph, P. F., Standardized complex and logarithmic eigensolutions for n-material wedges and junctions, *Int. J. Fracture*, Vol. 77 pp. 51–76, (1996).

Palaniswamy, K. and Knauss, W.G., *On the problem of crack extension in brittle solids under general loading*, Mechanics Today, ed. S. Nemat-Nasser, Pergamon Press., Vol. 4, (1978).

Palaniswamy, K. and Knauss, W. G., Propagation of a crack under general, in-plane tension, *Int. J. Fracture*, Vol. 8, pp. 114-117, (1972).

Pan E. A BEM analysis of fracture mechanics in 2-D anisotropic piezoelectric solids, *Engng Analy Boundary Elements.*, Vol. 23, pp. 67–76, (1999)

Pan, E. and Amadei, B., Boundary element analysis of fracture mechanics in anisotropic bimaterials. *Engineering Analysis with Boundary Elements*, Vol. 23, pp. 683-691, (1999).

Pan, E., A general boundary element analysis of 2-D linear elastic fracture mechanics. *Int. J. Fracture*, Vol. 88, pp. 41–59, (1997).

Pan, E. and Amadei, B., A 3-D boundary element formulation of anisotropic elasticity with gravity, *Appl. Math. Model.*, Vol. 20, pp. 114-120, (1996).

Pan, E. and Amadei, B., Fracture mechanics analysis of cracked 2-D anisotropic media with a new formulation of the boundary element method. *Int. J. Fracture*, Vol. 77, pp. 161-174, (1996).

Pan, E., *Gravitational and tectonic stresses in anisotropic rock masses with irregular topographies*, Ph.D. Thesis, Department of Civil, Environmental and Architectural Engineering, University of Colorado, (1993).

Peng, S. and Johnson, A. M., Crack growth and faulting in cylindrical specimens of Chelmsford granite, *International Journal of Rock Mechanics and Mining Sciences.*, Vol. 9, pp. 37-86, (1972).

Portela, A., Aliabadi, M.H. and Rooke, D.P., Dual boundary element incremental analysis of crack growth, *Composite Structures.*, Vol. 46, no.2, pp. 237-284, (1993).

Portela, A., *Dual boundary element analysis of crack growth*, Computational Mechanics Publications, Southampton UK and Boston USA, (1993).

Qu, J. Interface crack loaded by A time-harmonic plane wave, *Int. J. Solids & Stru.*, Vol. 31, pp. 329-345, (1996).

Rice, J. R., Elastic fracture mechanics concepts for interfacial crack, *Trans. ASME, J. Appl. Mech.*, Vol. 55, pp. 98-103 (1988).

Rice, J. R. and Sih, G. C., Plane problems of cracks in dissimilar media, *J. Appl. Mech.*, Vol. 32, pp. 418-423, (1965).

Richard, H. A., *Examination of brittle fractured criteria for overloapping mode I and mode II loading applied to cracks*, In Application of Fracture Mechanics to Materials and Structures. ed. Sih, G. C. *et al.* Martinus Nijhoff Pub., The Hague, (1984).

Rooke, D. P. and Tweed, J., The stress intensity factors of a radial crack in a point loaded disc, *Int. J. Eng. Sci.*, Vol. 11, pp. 285-290, (1973).

Sato, A., Hirakawa, Y. and Sugawara, K., Mixed mode crack propagation of homogenized cracks by the two-dimensional DDM analysis, *Construction and Building Materials*, Vol. 15, pp. 247-261, (2001).

Savage, W. Z., Amadei, B. and Swolfs, H. S., *Influence of rock fabric on gravity-induced stresses*, Proceed. of the Int. Symp. on Rock Stress and Rock Stress Measurements, Stockholm, Stephanson, O. (Ed.), Centek Publishers, Luleå, Sweden, (1986).

Scavia, C., A method for the study of crack propagation in rock structures, *Geotechnique*, Vol. 45, no. 3, pp. 447-463, (1995).

Shen, B. and Stephansson, O., Modification the G-criterion for crack propagation subjected to compression, *Engineering Fracture Mechanics*, Vol. 47, no. 2, pp. 177-189, (1994).

114

Sih, G. C. and Chen, E. P., Normal and shear impact of layered composite with a crack: dynamic stress intensification, *J. Appl. Mech.*, Vol. 47, pp. 351-358, (1980).

Sih, G. C., Strain-energy density factor applied to mixed mode crack problems, *Int. J. Fracture*, Vol. 10, No. 3, pp. 305-321, (1974).

Sih, G. C., *Handbook of stress intensity factors,* Inst. Of Fracture and Solid Mechanics, Lehigh University, Bethlehem, Pennsylvania, (1973).

Sih, G. C., Some basic problems in fracture mechanics and new concepts, *Eng. Fracture Mech.*, Vol. 5, pp. 365-377, (1973).

Sih, G. C., Paris, P. C. and Irwin, G. R., On cracks in rectilinearly anisotropic bodies, *Int. J. Fracture*, Vol. 3, pp. 189-203, (1965).

Sih, G. C., and Rice, J. R., Bending of plates of dissimilar materials with cracks, *J. Appl. Mech.*, Vol. 31, pp. 477-482, (1964).

Snyder, M. D. and Cruse, T. A., Boundary-integral equation analysis of cracked anisotropic plates, *Int. J. Fracture*, Vol. 11, pp. 315-328, (1975).

Sollero, P., Aliabadi, M. H. and Rooke, D. P., Anisotropic analysis of crack emanating from circular holes in composite laminates using the boundary element method, *Engineering Fracture Mechanics*, Vol. 49, no. 2, pp. 213-224, (1994).

Sollero, P. and Aliabadi, M. H., Fracture mechanics analysis of anisotropic plates by the boundary element method, *Int. J. Fracture*, Vol. 64, pp. 269-284, (1993).

Suo, Z., Singularities, interfaces and cracks in dissimilar anisotropic media, Proc. R. Soc. Lond. A, Vol. 427, pp. 331-358, (1990).

Swenson, D. O. and Rau Jr, C. A., The stress distribution around a crack perpendicular to an interface between materials, *Int. J. Fracture Mech.*, Vol. 6, pp. 357-365, (1970).

Tan, M.A. and Meguid, S.A., *Stress singularities of a sharp notch terminating at a bimaterial interface,* Proceedings of CANCAM 1997, June Quebec, (1997).

Tan, M. A. and Meguid, S. A., Dynamic analysis of cracks perpendicular to bimaterial interfaces using a new singular finite element, *Finite Elements in Analysis and Design.*, Vol. 22, pp. 69-83, (1996).

Timoshenko, S. and Goodier, J. N., *Theory of elasticity*, McGraw-Hill, New York, N. Y., (1970).

Ting, T. C. T., *Anisotropic elasticity: theory and application,* University Press, New York: Oxford, (1996).

Tippur, H. V. and Rosakis, A. J., Quasi-static and dynamic crack growth along bi-material interfaces: a note on crack-tip field measurements using coherent gradient sensing, *Experimental Mech.* Vol. 31, pp. 243-251, (1991).

Tsamasphyros, G. and Dimou, G., Gauss quadrature rules for finite part integrals, *Int. J. Numer. Methods Engng,* Vol. 30, pp. 13-26, (1990).

Vallejo, L. E., The brittle and ductile behavior of a material containing a crack under mixed-mode loading, *Proc. 28th U.S. Symp. Rock Mech.*, University of Arizona, Tucson, pp.383-390, (1987).

Wang, Q. Z., Jia, X. M., Kou, S. Q., Zhang, Z. X. and Lindqvist, P. A., More accurate stress intensity factor derived by finite element analysis for the ISRM suggested rock fracture toughness specimen CCNBD, *International Journal of Rock Mechanics and Mining Sciences,* Vol. 40, pp. 233-241, (2003).

Wang, W. C. and Chen, J. T., Theoretical and experimental re-examination of a crack perpendicular to and terminating at the bimaterial interface, *J. of Strain Analysis*, Vol. 28, pp. 53-61,(1993).

Westergaard, H., Bearing pressures and cracks, *Journal of Applied Mechanics,* Vol. 61, pp. 49-53, (1939).

Whittaker, B. N., Singh, R. N. and Sun, G., *Rock fracture mechanics : principles, design and applications.,* Elsevier, New York, (1992).

Wijeyewickrema, A.C., Dundurs, J. and Keer, L.M., The singular stress field of a crack terminating at a frictional interface between two materials, *J. Appl. Mech.*, Vol. 62, pp. 289-293, (1995).

Williams, M. L., The stresses around a faule or crack in dissimilar media, *Bulletin of Seismological Society of America*, Vol. 49, no. 2, pp. 199-204, (1959).

Williams, M. L., On the stress distribution at the base of a stationary crack, *J. Appl. Mech.* Vol. 24, pp.109-1 14, (1957).

Williams, M. L., Stress singularities resulting from various boundary conditions in angular corners of plates in extension, *J. Appl. Mech.*, Vol. 19, pp. 526-528, (1952).

Wong, R.H.C., Chau, K.T., Tang, C.A. and Lin, P., Analysis of crack coalescence in rock-like materials containing three flaws-Part I: experimental approach, *International Journal of Rock*

Mechanics and Mining Sciences, Vol. 38, pp. 909-924, (2001).

Woo, C. W. and Ling, H. L., On angle crack initiation under biaxial loading, *J. Strain Anal.*, Vol. 19, pp. 51-59, (1984).

Wu, C. H., Fracture under combined loads by maximum-energy-release rate criterion, *Journal of Applied Mechanics* Vol. 45, pp. 553-558, (1978).

Wu, K. C., Stress intensity factor and energy release rate for interfacial cracks between dissimilar anisotropic materials, *Trans. ASME, J. Appl. Mech.*, Vol. 57, pp. 882-886, (1990).

Wünsche, M., Zhang, Ch., Sladek, J., Sladek, V., Hirose, S. and Kuna, M., Transient dynamic analysis of interface cracks in layered anisotropic solids under impact loading, *Int.J. Fracture*, Vol. 157, no. 1-2, pp. 131-147, (2009).

Yang, H. J. and Bogy, D. B., Elastic wave scattering from an interface crack in a layered half-space, *J. Appl. Mech.*, Vol. 52, pp. 42-50, (1985).

Yuuki, R. and Cho, S. B., Efficient boundary element analysis of stress intensity factors for interface cracks in dissimilar materials, *Engineering Fracture Mechanics*, Vol. 34, pp. 179-188, (1989).

Zak, A. R. and Williams, M. L., Crack point stress singularities at a bi-material interface, *J. Appl. Mech.*, Vol. 30, pp. 142-143, (1963).

APPENDIX A

Experimental Results of Cement/Gypsum disc

Table A-1 CGD1

No.		CGD1-01	CGD1-02	CGD1-03	Ave.
Test	W_f(KN)	1.01	1.07	1.04	1.04
	Pre-crack length $2a$(mm)	23	24	22	23
	(β, λ)	0,90			
	(d_A, d_B, ξ_1, ξ_2)	0,0,90,94	0,0,89,88	0,0,92,92	0,0,90,91
	Initial angle $(\theta°)$ — Tip A	0	0	0	0
	Initial angle $(\theta°)$ — Tip B	0	0	0	0
Num.	Normalized SIFs of tip A — F_I	1.7645	1.7728	1.7573	1.7649
	Normalized SIFs of tip A — F_{II}	------	------	------	------
	Initial angle of tip A$(\theta°)$	0	0	0	0
	Normalized SIFs of tip B — F_I	0.7755	0.7930	0.7582	0.7756
	Normalized SIFs of tip B — F_{II}	------	------	------	------
	Initial angle of tip B $(\theta°)$	0	0	0	0

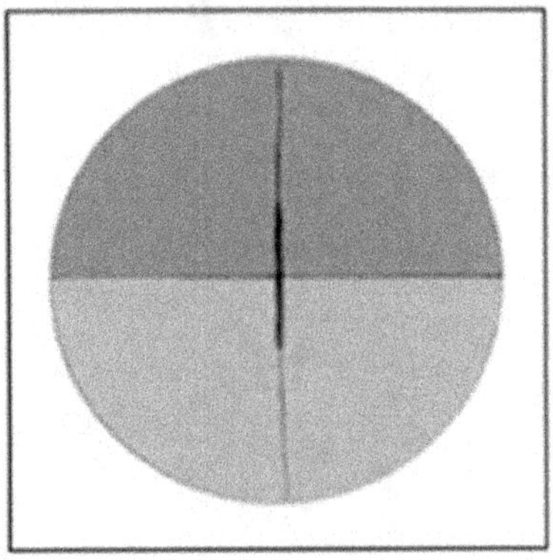

CGD1-01

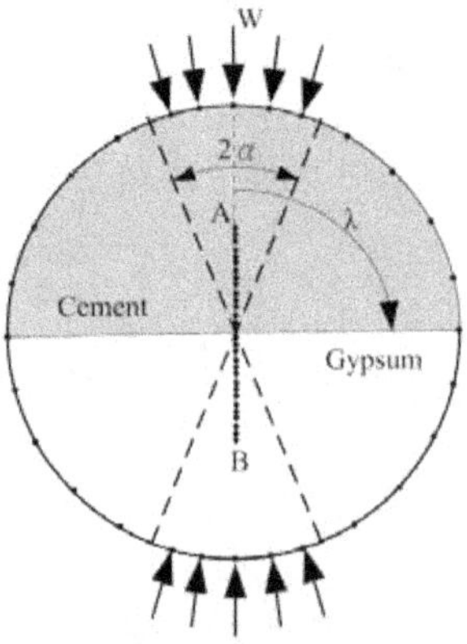

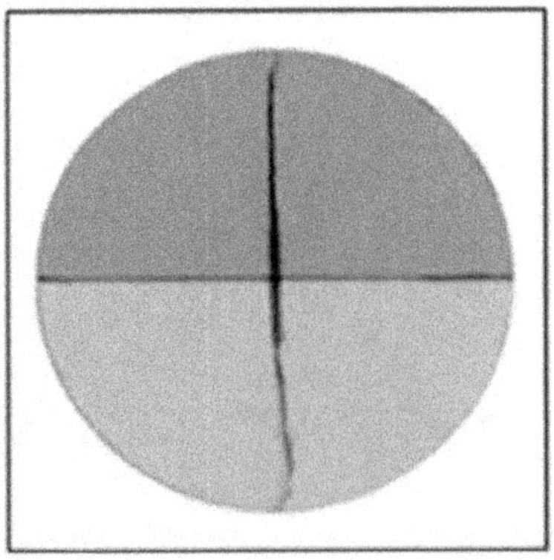

CGD1-02

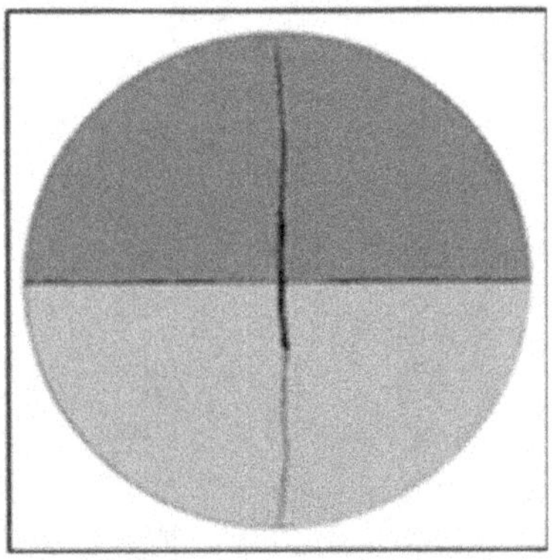

CGD1-03

118

Table A-2 CGD2

No.			CGD2-01	CGD1-02	CGD1-03	Ave.
Test	W_f(KN)		1.52	1.7	1.91	1.71
	Pre-crack length $2a$(mm)		21.6	20	20	20.5
	(β, λ)		(45,90)			
	(d_A, d_B, ξ_1, ξ_2)		1,1,88,92	1,1,96,92	1,1.5,90,89	1,1.2,91,91
	Initial angle $(\theta°)$	Tip A	77	77	78	77.33
		Tip B	86	87	88	87
Num	Normalized SIFs of tip A	F_I	-1.367	-1.324	-1.324	-1.338
		F_{II}	-2.890	-2.842	-2.842	-2.857
	Initial angle of tip A$(\theta°)$		78.772	78.389	78.389	78.517
	Normalized SIFs of tip B	F_I	-1.485	-1.478	-1.478	-1.481
		F_{II}	-1.699	-1.722	-1.722	-1.714
	Initial angle of tip B $(\theta°)$		87.578	87.278	87.278	87.378

CGD2-01

CGD2-02

CGD2-03

119

Table A-3 CGD3

No.			CGD3-01	CGD3-02	CGD3-03	Ave.
Test	W_f(KN)		1.054	1.057	1.048	1.053
	Pre-crack length $2a$(mm)		23	23	23	23
	(β, λ)		0,135			
	(d_A, d_B, ξ_1, ξ_2)		0,0,90,90	0,0,90,85	0,0,90,88	0,0,90,88
	Initial angle $(\theta°)$	Tip A	0	0	0	0
		Tip B	0	0	0	0
Num.	Normalized SIFs of tip A	F_I	-1.5127	-1.5127	-1.5127	-1.5127
		F_{II}	--			
	Initial angle of tip A$(\theta°)$		0	0	0	0
	Normalized SIFs of tip B	F_I	-0.722	-0.722	-0.722	-0.722
		F_{II}	--			
	Initial angle of tip B $(\theta°)$		0	0	0	0

CGD3-01

CGD3-02

CGD3-03

120

Table A-4 CGD4

No.			CGD4-01	CGD4-02	CGD4-03	Ave.
Test	W_f(KN)		1.87	1.99	2.08	1.98
	Pre-crack length 2a(mm)		23	24	22	23
	(β, λ)		45,135			
	(d_A, d_B, ξ_1, ξ_2)		1,1,92,85	1,0.5,90,84	0.5,1,90,92	0.8,0.8,91,87
	Initial angle ($\theta°$)	Tip A	59	60	58	59
		Tip B	57	56	65	59.33
Num	Normalized SIFs of tip A	F_I	-1.564	-1.598	-1.531	-1.564
		F_{II}	-2.561	-2.571	-2.550	-2.560
	Initial angle of tip A($\theta°$)		59.428	59.241	59.605	59.425
	Normalized SIFs of tip B	F_I	-1.008	-1.034	-0.982	-1.008
		F_{II}	-1.417	-1.416	-1.417	-1.416
	Initial angle of tip B ($\theta°$)		57.734	57.415	58.045	57.731

CGD4-01

CGD4-02

CGD4-03

121

Table A-5 CGD5

No.			CGD5-01	CGD5-02	CGD5-03	Ave.
Test	W_f(KN)		1.36	1.56	1.79	1.57
	Pre-crack length 2a(mm)		23	24	22	23
	(β, λ)		45,45			
	(d_A, d_B, ξ_1, ξ_2)		0,0,90,91	0,0,92,95	0,0,88,90	0,0,90,92
	Initial angle $(\theta°)$	Tip A	78	83	80	80.33
		Tip B	90	82	85	85.67
Num	Normalized SIFs of tip A	F_I	-1.259	-1.248	-1.265	-1.257
		F_{II}	-1.795	-1.794	-1.794	-1.794
	Initial angle of tip A$(\theta°)$		81.23	82.34	80.54	81.37
	Normalized SIFs of tip B	F_I	-1.595	-1.575	-1.615	-1.595
		F_{II}	-1.830	-1.828	-1.833	-1.830
	Initial angle of tip B $(\theta°)$		88.15	89.03	87.51	88.23

CGD5-01

CGD5-02

CGD5-03

122

Table A-6 CGD6

No.		CGD6-01	CGD6-02	CGD6-03	Ave.
Test	W_f(KN)	0.99	0.92	1.02	0.98
	Pre-crack length $2a$(mm)	22	23	22	22.33
	(β, λ)	0,0			
	(d_A, d_B, ξ_1, ξ_2)	0,0,91,89	0,0,91,89	0,0,90,90	0,0,91,89
	Initial angle ($\theta°$) — Tip A	0	0	0	0
	Initial angle ($\theta°$) — Tip B	0	0	0	0
Num.	Normalized SIFs of tip A — F_I	0.9622	0.9716	0.9622	0.9653
	Normalized SIFs of tip A — F_{II}	---			
	Initial angle of tip A($\theta°$)	0	0	0	0
	Normalized SIFs of tip B — F_I	0.9804	0.9904	0.9804	0.9837
	Normalized SIFs of tip B — F_{II}	---			
	Initial angle of tip B ($\theta°$)	0	0	0	0

CGD6-01

CGD6-02

CGD6-03

123

Table A-7 CGD7

No.			CGD7-01	CGD7-02	CGD7-03	Ave.
Test	W_f(KN)		1.75	0.92(x)	1.83	1.79
	Pre-crack length $2a$(mm)		22	20.2	22	21.4
	(β, λ)		45,0			
	(d_A, d_B, ξ_1, ξ_2)		1,0,85,88	0,10,105,92	0,0,85,88	0.5,0,85,88
	Initial angle $(\theta°)$	Tip A	82	135	83	82.5
		Tip B	70	0	69	69.5
Num	Normalized SIFs of tip A	F_I	-0.674	-0.683	-0.674	-0.677
		F_{II}	-1.336	-1.632	-1.336	-1.345
	Initial angle of tip A$(\theta°)$		82.08	80.54	82.08	81.57
	Normalized SIFs of tip B	F_I	-2.420	-2.419	-2.420	-2.421
		F_{II}	-2.805	-2.810	-2.805	-2.807
	Initial angle of tip B $(\theta°)$		67.87	67.76	67.87	67.83

CGD7-01

CGD7-02

CGD7-03

124

www.ingramcontent.com/pod-product-compliance
Lightning Source LLC
Chambersburg PA
CBHW081131170526
45165CB00008B/2630